DESIGN COMMUNICATION

BLT
BULLETI
DAVID PAUL HELPERN, PC
drives those
past those wh
LEA GROUP UPDATE
Al Cohen Co
Good space is good business.
Business Space Design
iNSiDE!
Space Planning
Questionnaire
Greiner Engineering
LAND DEVELOPMENT
ARCHITECTURE
Fees.
Budgets.
Team.
Quick.
JFMAMJJASOND
WORLD EXPOSITION 1984
A PERFECT BLEND
Happy Thanksgiving
A Turkey-graph
BOBROW THOMAS ASSOCIATES
Anniversary
OPEN HOUSE
Seattle University
1986
PARTNERSH

DESIGN COMMUNICATION

Developing Promotional Material For Design Professionals

Ernest Burden

McGraw-Hill Book Company

New York St. Louis San Francisco Auckland
Bogota Hamburg London Madrid Mexico Milan
Montreal New Delhi Panama Paris Sao Paulo
Singapore Sydney Tokyo Toronto

Library of Congress Cataloging-in-Publication Data

Burden, Ernest E., 1934-
  Design Communication.

  Includes Index.
  1. Architectural design-Audio-visual aids.
2. Communication in architectural design. 3. Design,
Industrial--Audio-visual aids. 4. Design--Audio-visual
aids. 5. Communication in design. I. Title.
NA2750.B85 1987   729'.068'8   87-10040
ISBN 0-07-008932-9

ISBN 0-07-008932-9

Copyright © 1987 by McGraw-Hill, Inc. All rights reserved.
Printed in the United States of America. Except as permitted
under the United States Copyright Act of 1976, no part of this
publication may be reproduced or distributed in any form or by
any means, or stored in a data base or retrieval system, without
the prior written permission of the publisher.

1234567890 HAL / HAL 89210987

CONTENTS

Acknowledgments IV

Preface V

1: Communications Programs 1

2: Designing The Format 18

3: Direct Mail 46

4: Special Events 70

5: Company Brochures 90

6: Case Studies 146

7: Leasing Brochures 200

Design Credits 212

Index 216

ACKNOWLEDGMENTS

The production of a book as complex as this one requires the combined talents of many people all the way from concept to completion. In design, editorial, production and printing, everyone involved has exhibited their skills to the utmost.

Beyond those associated with the publishing and printing of this work, there was an extraordinary effort put in by the people who assisted me in the production of this book. I am deeply grateful to them for what seemed at times to be an endless task. To Laura Sutcliffe I owe the most thanks for staying with this project from start to finish and for overseeing the entire production, and to Joy Arnold for her assistance in the final mechanicals and for quality control. The Specific credits are as follows:

CONCEPTUAL DESIGN / EDITING: By the Author
COVER DESIGN: Laura Sutcliffe
COVER PHOTO: Litchfield and Barzda
PAGE LAYOUTS: Laura Sutcliffe
PROJECT DESCRIPTIONS: Madeleine Beckman/Laura Sutcliffe
EDITING AND TYPESETTING: Laura Sutcliffe
PHOTOS AND MECHANICAL PREPARATION: Laura Sutcliffe
FINAL MECHANICALS/QUALITY CONTROL: Joy Arnold

PREFACE

Communications is a vital function of a marketing program within design firms today. While a communications program supports the marketing efforts, it does not replace them; it enhances them. It gives marketing the tools it needs. But more important, it gives design firms the freedom to do their routine work while the promotional material is doing its job of making new contacts and getting new work.

The promotional material is helping to pre-sell the firm in a positive manner. Nothing else can do that except for personal contact. We have to be present with our clients to verbalize our thoughts, but promotional tools can work for us as a silent sales force. The promotional tools get the invitation to make the personal call.

The challange is to bring the three-dimensional reality of projects, people and a firm's experience into a visual format that can be presented to others. Firms need to package their three-dimensional design products in two-dimensional media either for mass distribution or to a targeted audience.

You can take a client to see your completed projects but this is time consuming, expensive and in the case of busy clients, difficult or impossible to schedule. Therefore the promotional material must be a close simulation of reality of being there on the site and of talking with your people and learning about your expertise.

Before the marketing pieces are designed, it is important to know what the marketplace wants in general and what selected clients in particular want to know about your firm. With an objective evaluation of your prospective client's needs, the communication process becomes a positive activity insted of a guessing game.

The projects selected for this book illustrate a wide variety of positive approaches taken by design firms nationwide. They are presented in a case study format, so the dominant features of each selection can be illustrated and described on their own merits. There are no design judgments made in selecting materials to be included, although many pieces shown are award winners.

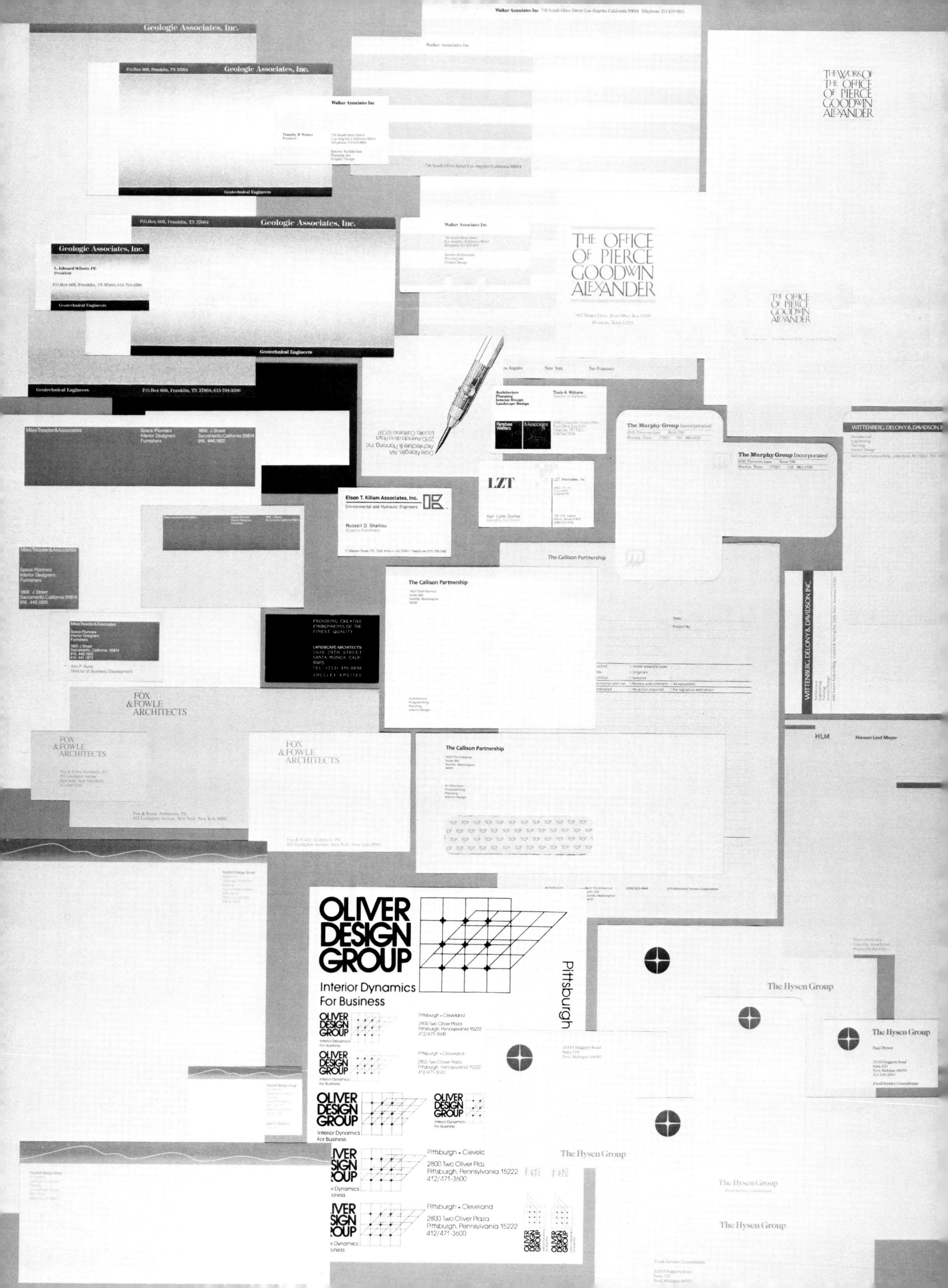
Geologic Associates, Inc.
P.O.Box 668, Franklin, TN 37064
Geotechnical Engineers
L. Edward Wilson, PE
President
P.O.Box 668, Franklin, TN 37064; 615-794-3596
Walker Associates Inc 716 South Olive Street Los Angeles California 90014 Telephone 213 629 0011
Walker Associates Inc
Timothy B Walker
President
716 South Olive Street Los Angeles California 90014
THE WORK OF THE OFFICE OF PIERCE GOODWIN ALEXANDER
THE OFFICE OF PIERCE GOODWIN ALEXANDER
Los Angeles
New York
San Francisco
Dale Naegle, AIA
Architecture & Planning Inc
2210 Avenida de la Playa
La Jolla, California 92037
Architecture
Planning
Interior Design
Landscape Design
Fersboe Walters & Associates
Truely A. Williams
Director of Marketing
The Murphy Group Incorporated
Houston, Texas 77027
WITTENBERG, DELONY & DAVIDSON, INC.
Miles Treaster & Associates
Space Planners
Interior Designers
Furnishers
1800 J Street
Sacramento, California 95814
916. 446.1800
Elson T. Killam Associates, Inc.
Environmental and Hydraulic Engineers
Russell D. Shaillieu
Graphics Coordinator
LZT
Ken Lynn Durfee
LZT Associates, Inc.
Ann P. Ayre
Director of Business Development
PROVIDING CREATIVE ENVIRONMENTS OF THE FINEST QUALITY
LANDSCAPE ARCHITECTS
2646 29TH STREET
SANTA MONICA, CALIF. 90405
TEL (213) 450-8898
SHELLEY KNEITEL
The Callison Partnership
1423 Third Avenue
Suite 300
Seattle, Washington
98101
Architecture
Programming
Planning
Interior Design
Date:
Project No:
Under separate cover
Originals
Samples
Review and comment
As requested
No action required
For signature and return
FOX & FOWLE ARCHITECTS
Fox & Fowle Architects, P.C.
192 Lexington Avenue, New York, New York 10016
HLM
Hansen Lind Meyer
A Professional Service Corporation
OLIVER DESIGN GROUP
Interior Dynamics For Business
Pittsburgh
Pittsburgh • Cleveland
2800 Two Oliver Plaza
Pittsburgh, Pennsylvania 15222
412/471-3600
The Hysen Group
Paul Hysen
Food Service Consultants
Novi, Michigan 48050

COMMUNICATIONS PROGRAMS: 1

Corporate Communications

How To Develop A Corporate Communications Program

There is a definite continuous relationship between the marketing activities of a design firm and the delivery of the project. This relationship is evident throughout the life of a project. Project delivery should begin when client objectives are defined and analyzed. Similarly, the marketing effort should proceed until the project is completed and continue after the project is completed. The coordination and integration of these two functions form one of the key factors in the success or failure of the design firm. The diagram at the right outlines this process.

Goals And Objectives

The overall goals of the firm should be identified in terms of its potential growth within a market or markets, new markets to explore potential income growth, and an increase in the number of personnel required to sustain that growth. The immediate challenges of any firm are to identify and clarify the targeted clients within each market, to develop a marketing plan to address the overall direction of the firm, and to specify a direction for each market segment. Two goals that marketing plans address are: to increase awarness among past, current and potential clients in the firm's markets, and to instill pride in the firm among its employees and prospective employees.

The Marketing Plan

The written marketing plan is the guide behind the marketing communications, in terms of research, design and production. There should also be a written communications plan that lists specific collateral material for promotional use. Once a firm has identified the markets to pursue, the plan identifies how to reach those markets effectively with the appropriate material and message. It also identifies what format to use and what should be highlighted in each piece of marketing material. Each item in the marketing plan is given a priority and specific relationship to all the others.

Communications Plan

The communications plan should contain a schedule and a budget for all items to be developed. The schedule should outline when all the pieces are to be prepared and how long it will take to produce them. An integral part of the schedule is the budget. The communications budget does not include marketing salaries but should include public relations retainers, advertising, and funds for writers, photographers and graphic artists. It should also include production, printing and mailing costs of all items. A comprehensive corporate communications plan includes, among other things, a corporate identity program, an advertising and public relations program, and a direct selling campaign.

Corporate Identity

A corporate identity program uses a graphic representation of the firm to a variety of publics which fulfills all operational and marketing needs. This graphic image is applied to the entire package of print and visual materials used by the firm. The first visual contact a client may have with any firm is its logo and other business items such as letterheads. The logo should be appropriate to the majority of clients served. The communications program should creatively express the firm's operation, display its unique character and capabilities, and provide a workable format for the design of collateral promotional material.

Advertising And Publicity

The second element of a corporate communications program consists of individual advertisements or campaigns placed in business, trade or general-interest publications. Another form of general public contact is where material is developed for display and dissemination at regional or national professional or market-based trade shows. Here a firm can market its capabilities, or develop visibility, or introduce a specific service. Publicity and public relations programs previously covered everything a firm did to promote itself. Today promotion is more highly specialized and brings awareness to potential clients through many diverse items such as organized tours, seminars and lectures.

Direct Mail Programs

The third element of a corporate communications program involves direct-mail programs, wherein marketing materials are developed to reach a specific target audience and to elicit a response. A direct mail campaign can include form letters, pamphlets, response mechanisms, announcements, and market-specific brochures. A direct-mail campaign gets specific messages to targeted audiences. There is a wide range of items that can be used as direct-mail items, such as brochures and fliers, newsletters and magazines, posters and announcements and reprints of articles from trade publications.

Special Events Pieces

Two items round out the promotional materials list, one at each end of the spectrum. The special events piece usually represents a one-time effort for a special occasion or event. This can include anniversaries, a new name, new location of office, staff promotion, or client oriented announcement, such as groundbreakings and dedications. On the other end of the scale is the annual report. These are for distribution externally for publicly held or privately owned companies. They usually focus on accomplishments of the firm during the past year, progress, future expectations and a financial report.

COMMUNICATION TOOLS

Although there are many communication tools, the most common way of showing the experience of a firm is the corporate brochure. It is the firm's number-one marketing tool, although it may be augmented with specialty brochures or special market brochures. These represent efforts to market a particular service, discipline or branch office to a targeted market segment.

Another form of direct-mail promotion is through newsletters and magazines. Newsletters are usually a high-frequency external publication with few pages, employing a concise writing style. Generally the style is a news or feature format. Magazines are external publications with a liberal use of photos and art and usually use a more interpretive writing style. The format is generally feature oriented.

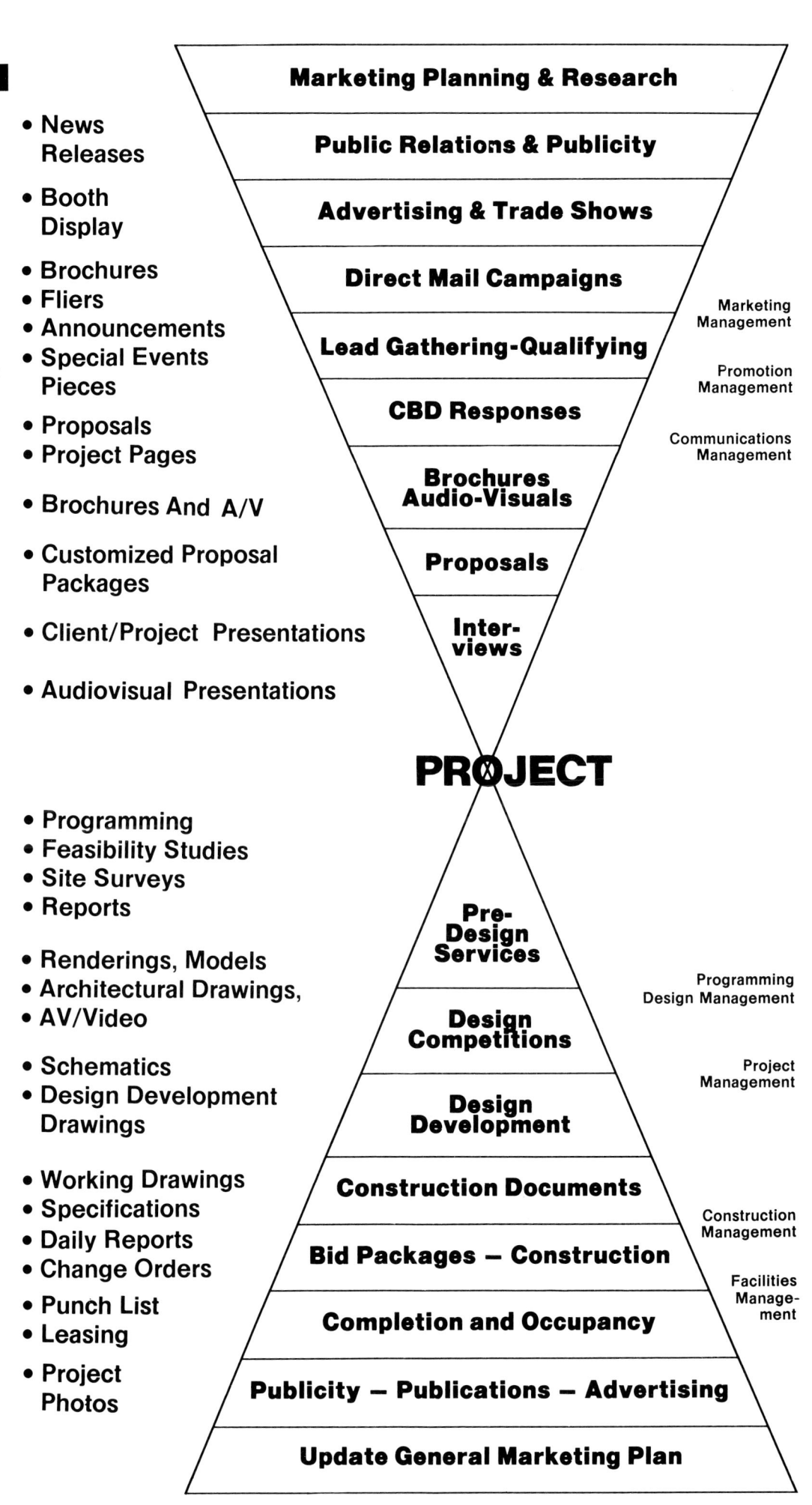

- News Releases
- Booth Display
- Brochures
- Fliers
- Announcements
- Special Events Pieces
- Proposals
- Project Pages
- Brochures And A/V
- Customized Proposal Packages
- Client/Project Presentations
- Audiovisual Presentations

- Programming
- Feasibility Studies
- Site Surveys
- Reports

- Renderings, Models
- Architectural Drawings,
- AV/Video

- Schematics
- Design Development Drawings

- Working Drawings
- Specifications
- Daily Reports
- Change Orders
- Punch List
- Leasing
- Project Photos

Marketing Planning & Research
Public Relations & Publicity
Advertising & Trade Shows
Direct Mail Campaigns
Lead Gathering-Qualifying
CBD Responses
Brochures Audio-Visuals
Proposals
Interviews

PROJECT

Pre-Design Services
Design Competitions
Design Development
Construction Documents
Bid Packages — Construction
Completion and Occupancy
Publicity — Publications — Advertising
Update General Marketing Plan

Marketing Management
Promotion Management
Communications Management

Programming Design Management
Project Management
Construction Management
Facilities Management

Monograms

A monogram is a standardized typographic treatment derived from the first initials of a person's name. The use of the monogram is widespread in almost every industry, and it can be particularly useful for design firms and consultants. Design firms usually use their individual name, or a group of partners' names, for the corporate identity of the firm. By using the monogram in a clear concise manner, firm recognition is strengthened. The monogram can be a useful graphic tool for unifying a firm's identity. The graphic design of a monogram can be a completely literal combination of letters in a particular type style, or a more abstract treatment, which approaches a logo or symbol.

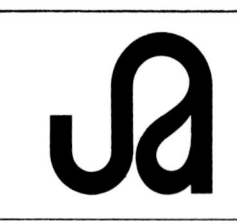

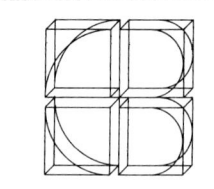

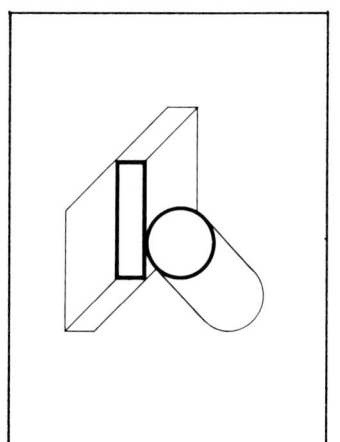

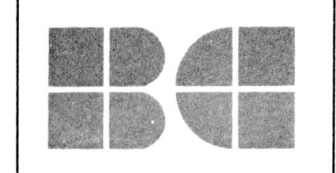

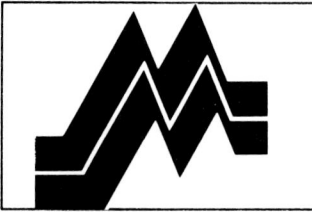

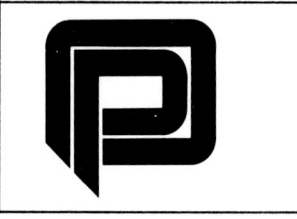

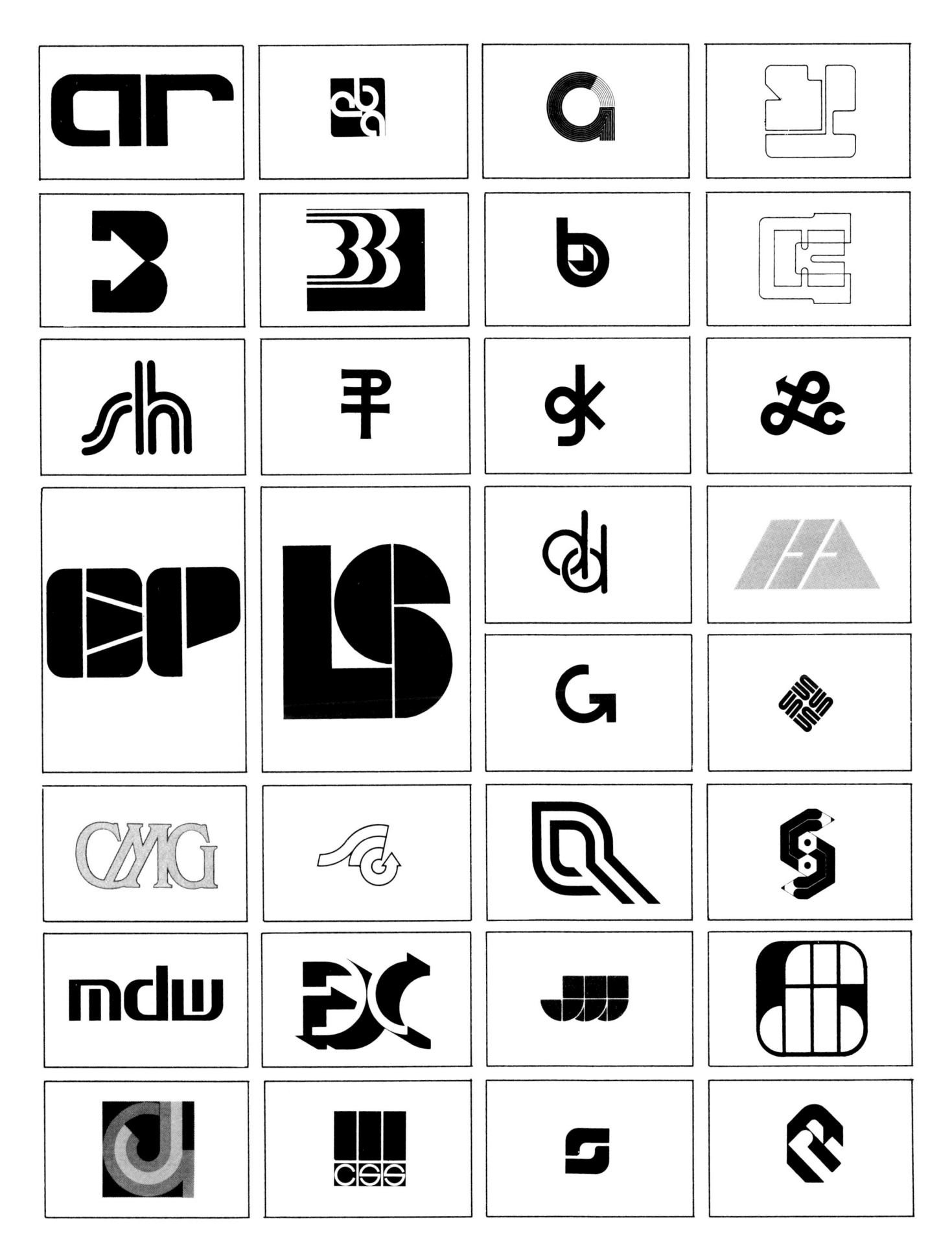

Communications Programs: 5

Logos

A commonly used identity symbol for design firms, as well as industry in general, is the logo. It is a compact graphic shape used to represent an image of the organization. A logo can stand alone or can be used in conjunction with a type face. A logo can be very effective in creating a strong sense of unity and recognition, when given repeated exposure. The logo may appear as an abstract design or as a graphic representation of the firm's main discipline. The logo should be used to communicate the firm's best qualities to the prospective client through a clear graphic representation. The logos shown here are representative of design firms.

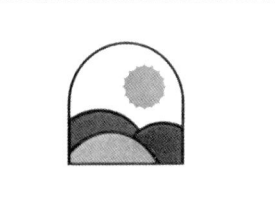

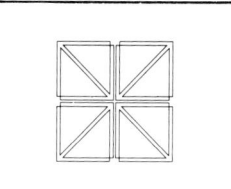

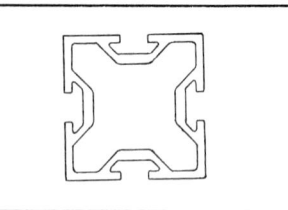

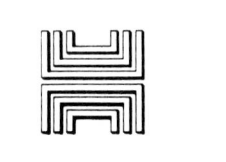

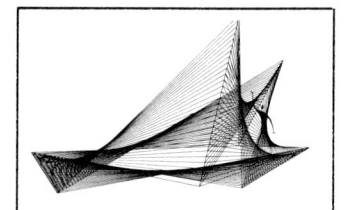

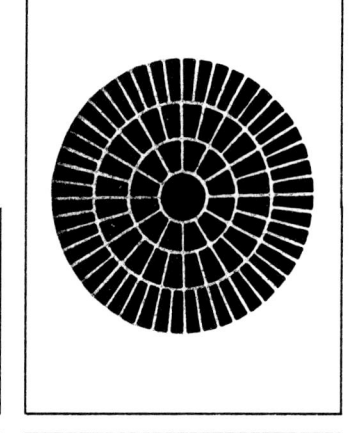

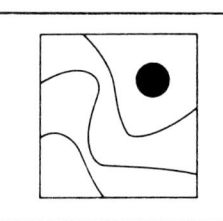

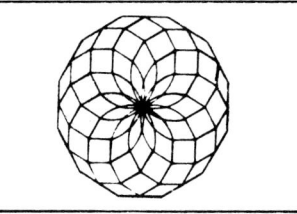

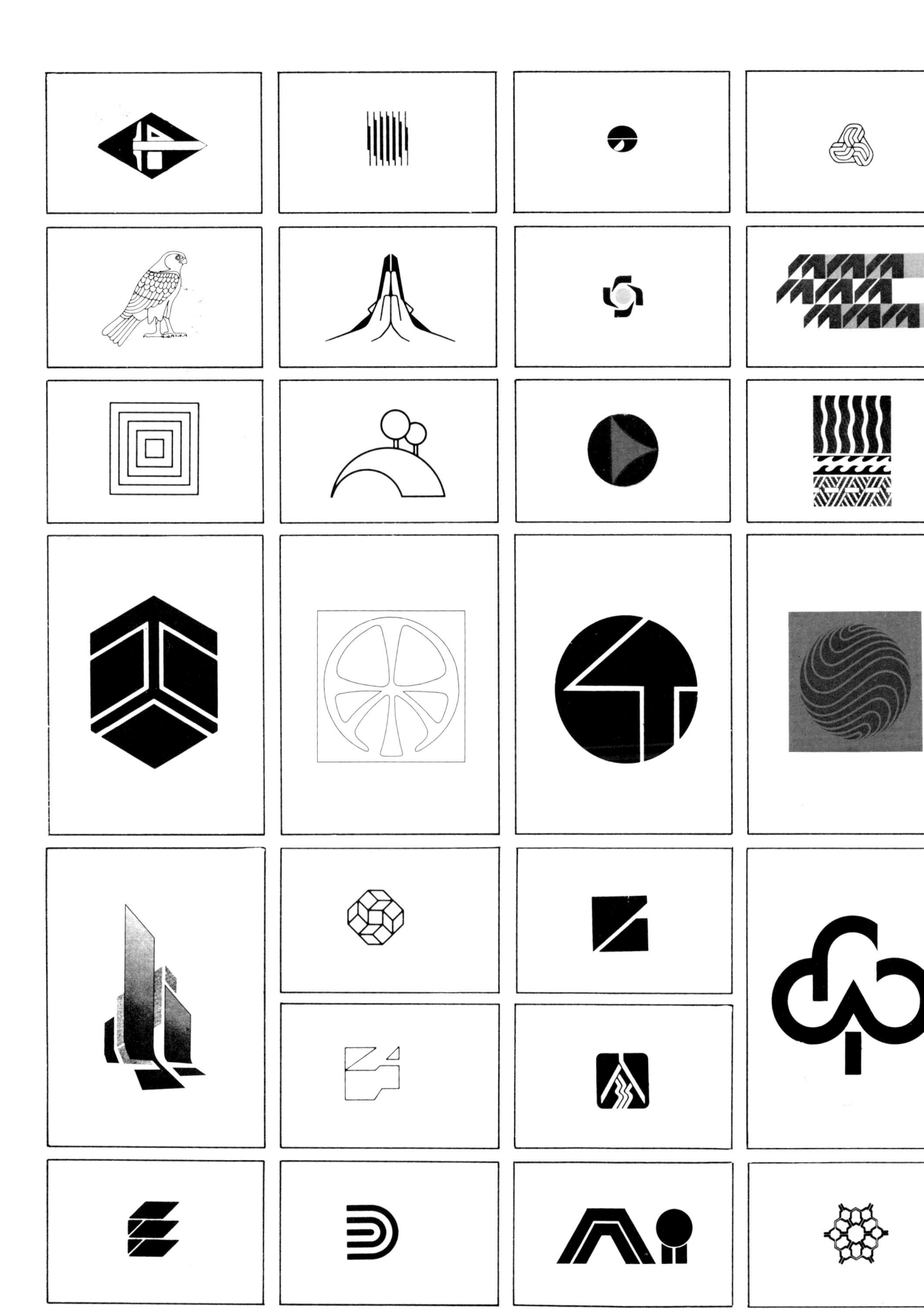

Communications Programs: 7

Logotypes

Logotypes, which are usually used in conjunction with a logo or monogram, can also be used independently. The selection of a suitable typeface, the placement of the firm's name and the design decision to use or not to use any graphic elements, such as bars or rules, can make each logotype as unique as a logo itself. Since most typefaces create a specific image, the selection of an appropriate typeface is important. Typefaces can also be modified, thus creating a more distinct identification for the firm.

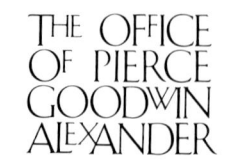

GEORGE THOMAS HOWARD ASSOCIATES

THE KLING PARTNERSHIP

Craig & Lawson Architects

WALLACE WENDEL ARCHITECT • AIA

CAPELIN COMMUNICATIONS, INC.

Turner Collie & Braden Inc.

GRUZEN & PARTNERS

GATHMAN · MATOTAN ARCHITECTS · PLANNERS INC.

ALFRED T. KUREK ASSOCIATES Architects

Gelsomino · Johnson Architects

DeREVERE WISE GARAKIAN AND ASSOCIATES
ARCHITECTURE / PLANNING

THE OFFICE OF BAHR VERMEER & HAECKER ARCHITECTS
1030 QUE STREET
LINCOLN, NEBRASKA 68508

MEININGER

ARIX

PANZICA

BOWERS

Holabird & Root

bowyer-singleton & associates
INCORPORATED

McGILL & SMITH

Perkins & Will

Communications Programs: 9

Symbols and Logotypes

Most symbols (or logos) are accompanied by the firm's full name in a stylized letter font called a logotype. Using the two in combination serves the purpose of introducing the intended audience to a logo which hopefully will become synonomous with the firm's full name. The typeface chosen to accompany the logo should reinforce the logo's design, giving the impression of a solid graphic unit. This unit may or may not include the full amount of information, such as discipline(s), address and phone number.

ARVID ELNESS **ARCHITECTS** INC.
200 BUTLER NORTH • 510 1ST AVE. N.
MINNEAPOLIS, MINNESOTA 55403

TRIAD ASSOCIATES

VECO
engineers architects surveyors

Clovis Heimsath Associates AIA
Architects

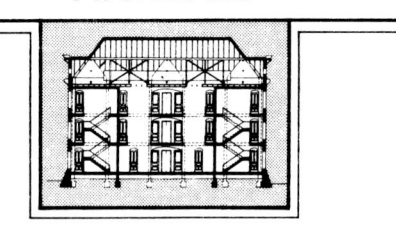

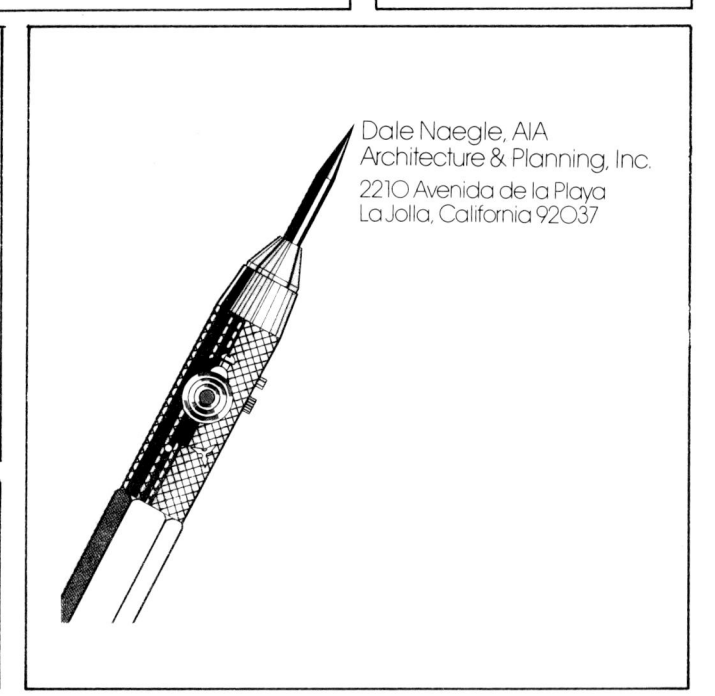

CH2M
HILL
engineers
planners
economists
scientists

DESIGN WEST
ARCHITECTURAL DESIGN WEST INC.
ARCHITECTS ENGINEERS PLANNERS
5155 NORTH FIVE MILE ROAD
BOISE IDAHO 83702   TEL 208-376-9200

bb
BRIAN BURR
ARCHITECTURAL
ILLUSTRATOR
343 EAST 30TH STREET
NEW YORK, N.Y. 10016
PHONE (212) 689 9387

Luckie
&
Forney
Inc.
Advertising

Dale Naegle, AIA
Architecture & Planning, Inc.
2210 Avenida de la Playa
La Jolla, California 92037

Ayres, Lewis, Norris & May, Inc.
Engineers · Architects · Planners

LOUIS G. REDSTONE ASSOCIATES, INC.   Architecture/Engineering/Planning/Interior Design

BONSAL

10

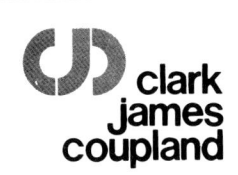

Monograms and Logotypes

If a firm is identified solely by its initials, such as HLW or RVBK, the use of a combination monogram and logotype may be necessary for identification purposes. A firm that uses a monogram for design reasons, doubling as a logo, will find the combination of monogram and logotype ties the firm name to a certain graphic look and symbol. The type face chosen for the logotype should complement the monogram.

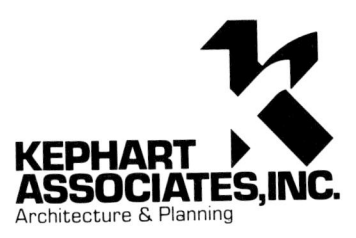

Communications Programs: 11

Corporate Identification Manual

Corporate identity manuals are created to establish guidelines for the proper use and placement of the firm's corporate identity symbols on various pieces of collateral and business forms. Consistent and correct adherence to the rules laid down in these manuals is important to the success of the program as a whole. The size of the manual may vary from a few printed pages to a large volume depending on the concern of the firm for specific controls of its identification program.

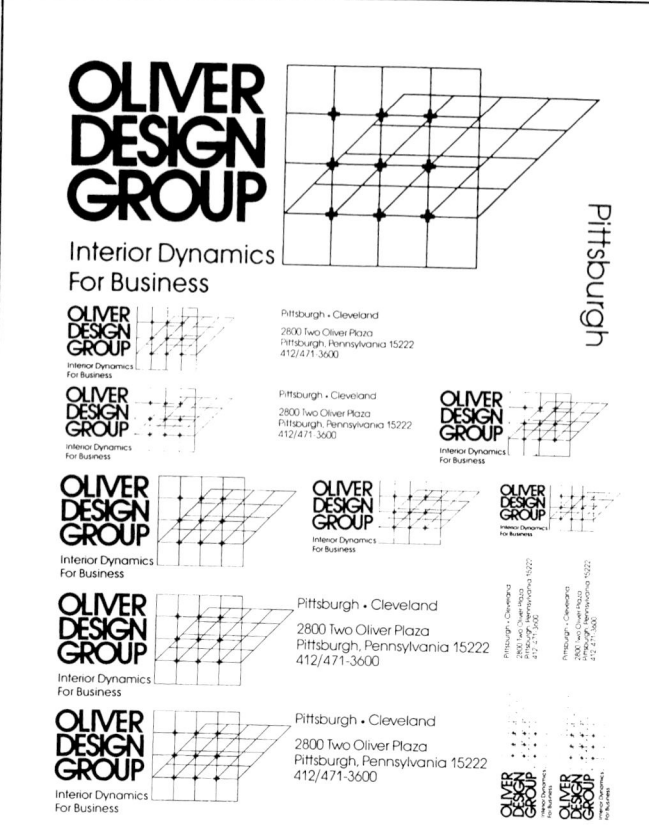

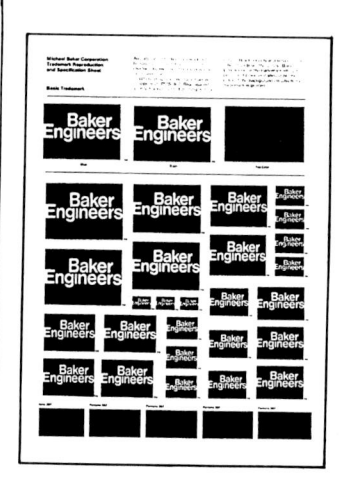

Some of the applications typically illustrated in corporate identification manuals are:

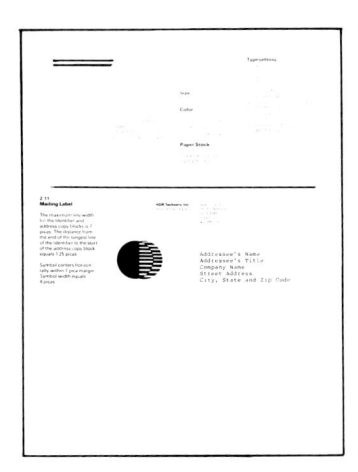

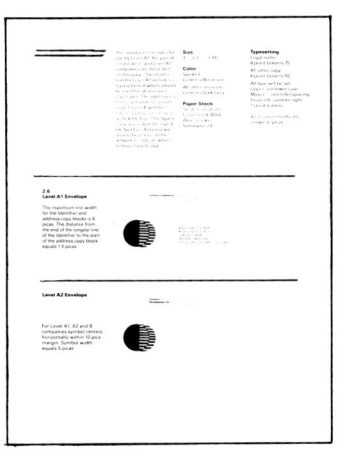

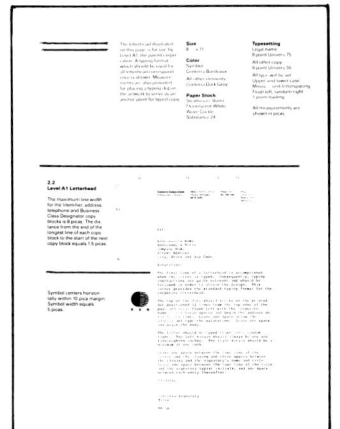

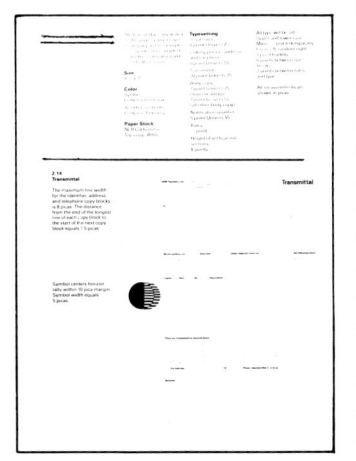

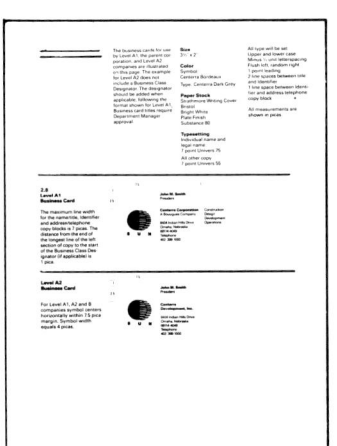

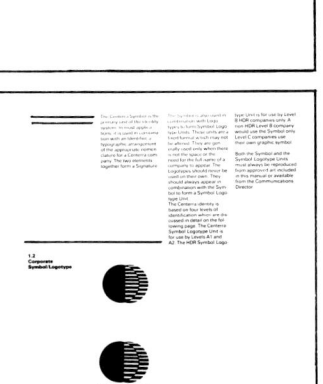

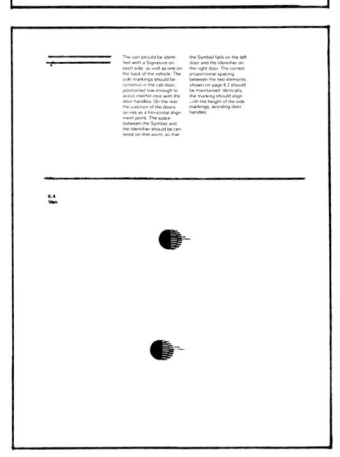

BASIC STANDARDS
- Glossary of Terms
- Corporate Symbol/ Grid/Guidelines
- Levels of Identification
- Signature Configurations
- Design Control
- Color Standards / Swatches
- Decorative Use
- Reverse Use
- Supporting Typography

STATIONERY SYSTEM
- Introduction
- Letterhead
- Envelope
- Business Card
- Personalized Note Pad
- Mailing Label
- Press Release
- Internal Memorandum
- Transmittal
- Proposal Covers
- Project Report
- Title Page Format
- Purchase Order
- Invoice
- Check/Envelope

ADVERTISING AND PROMOTION
- Newsletter
- Brochures
- Binders
- Image Ads
- Calling Card Ads
- Yellow Page Ads

SIGNAGE SYSTEM
- Exterior Formats
- Interior Formats
- Construction Sign Formats

VEHICLE SYSTEM
- Trucks, Vans, Cars
- Trailer
- Corporate Jet

TITLE BLOCKS
- Cover Sheet 8½x11
- Title Blocks

BASIC ITEMS
- Hard Hat/Name Tag
- Coaster/Lapel Pin/ Napkin/Cup

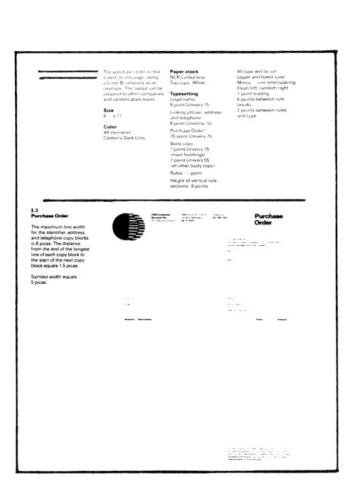

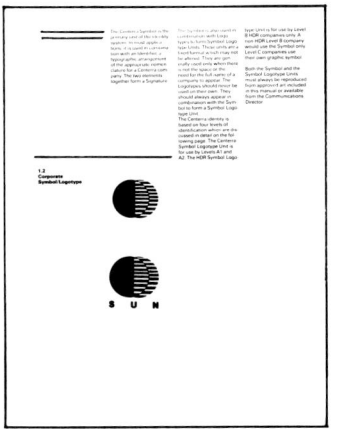

Communications Program 13

Developing
A Corporate
Identity

Logos and logotypes are important images of a firm and are not changed at a whim, or redesigned overnight. In the case of this multi-disciplined firm one of the three partners retired and a study was made to redesign the logo and logotype. A series of graphic design studies and alternatives were presented before arriving at a solution. For example, the combination of the firm's partners' names prompted many studies for the logo. One of the names led to the study of the dewberry shaped logo. At the same time the corporate color, burgandy, was chosen representing the color of the dewberry.

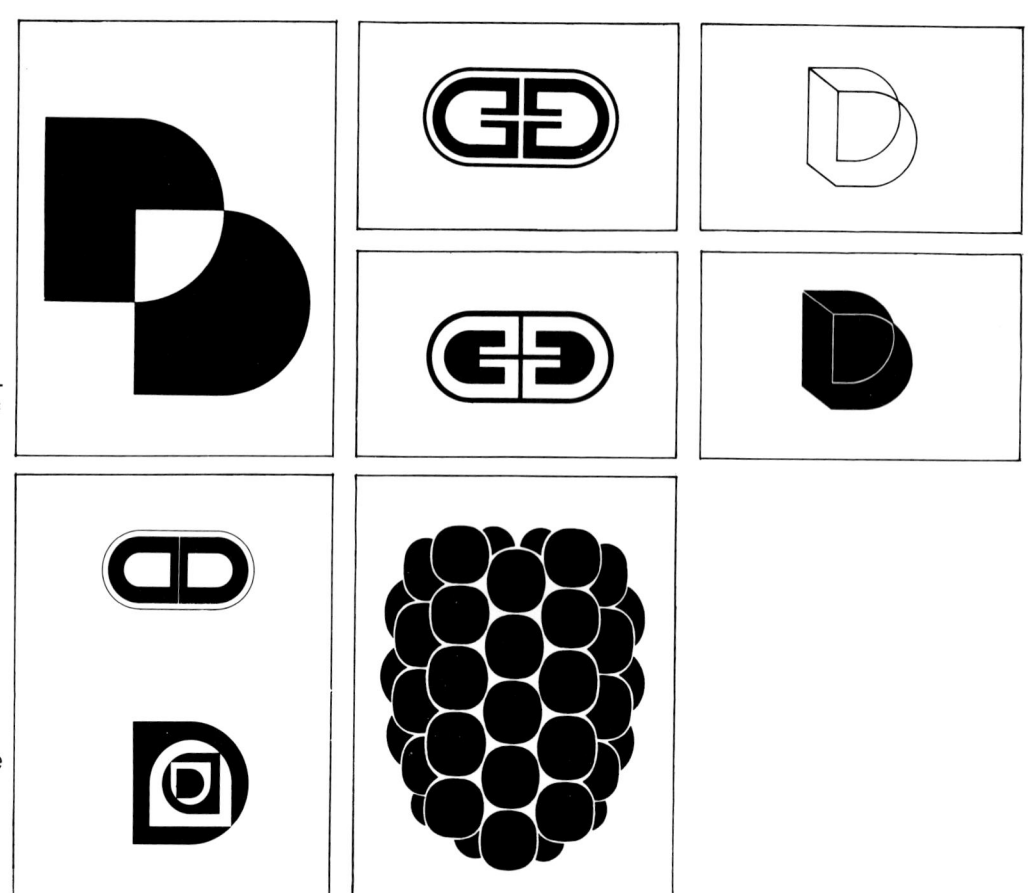

Various studies were made to actual size of all of the logos that were being considered. The placement on the page was studied in a typeset mock-up. The logotype was studied in conjunction with the selected finalist designs for the logo. For example, there are several ways that two names can be combined, one way is with an "and", another way is with a plus (+), or an ampersand (&). The firm chose the ampersand and the dewberry symbol as the final selection.

Dewberry/Davis Associates

Dewberry/Davis Associates

Dewberry and Davis

Dewberry and **Davis**

Dewberry+Davis

Dewberry & Davis

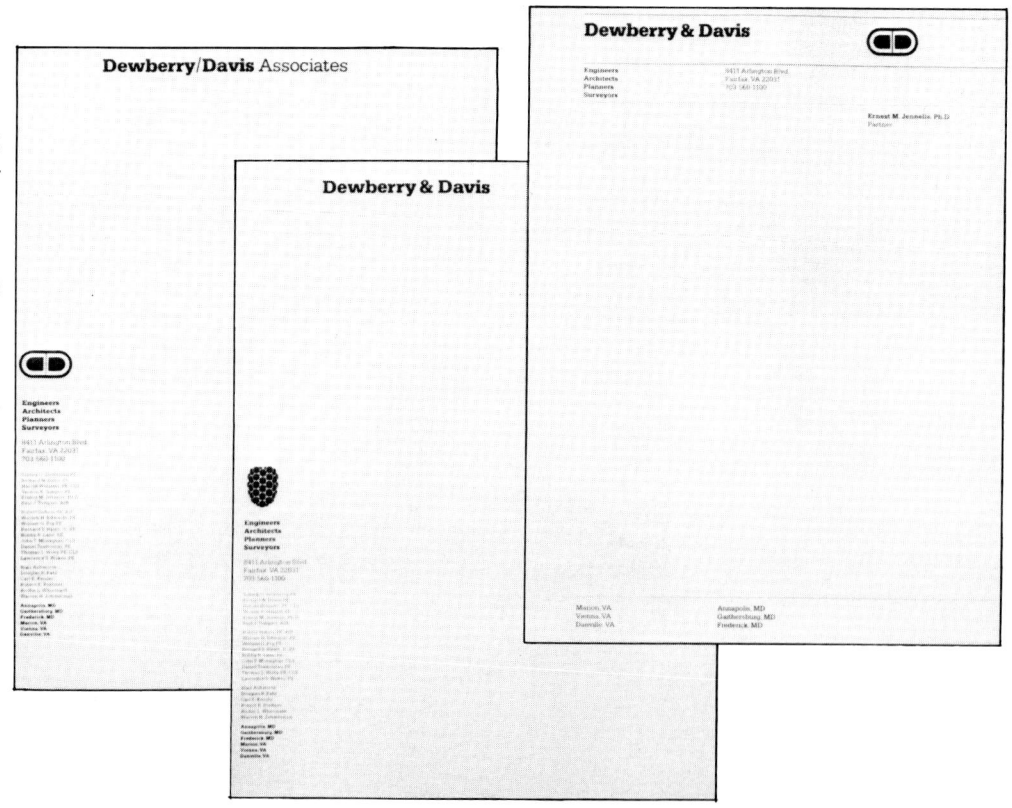

14

Dewberry & Davis

Engineers
Architects
Planners
Surveyors

The logo was then studied for various other promotional pieces such as project pages, the firm's newsletter and a corporate brochure. The decision was made to place the logo on the bottom of all of these pieces.

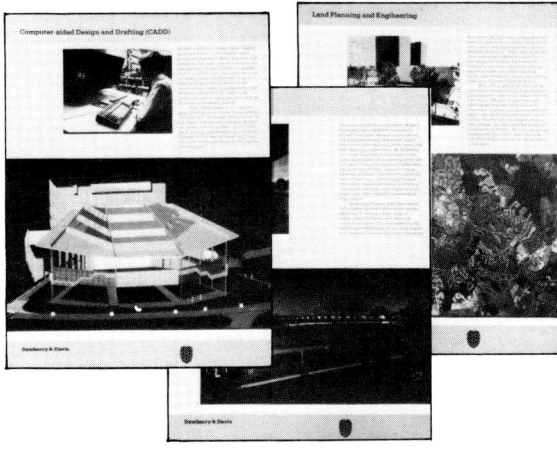

In addition to the burgandy color of the logo, a silver, gray, and black color scheme was developed for the other pieces. Most of the photographs were set into a matt black field with a silver band at the top of the page. Rules appeared under each photograph in the burgandy color.

Communications Programs 15

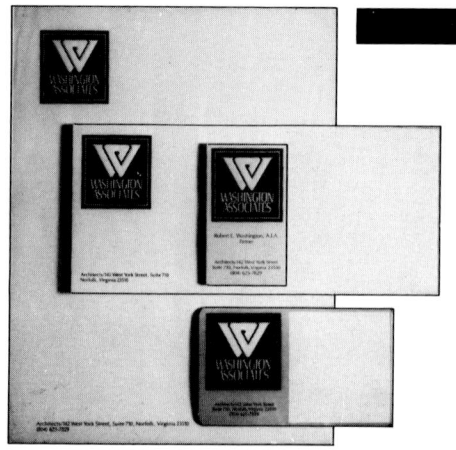

Washington Associates

This architectural firm has designed a distinctive logo for all its corporate ID and promotional pieces. The logo consists of a large stylized "W" and the firm's name printed with white type on a black background. While the "W" and the firm name are contained in this black box, all other pertinent information (discipline, address and phone number) is placed outside the box.

The firm uses its logo as the only graphic on the cover of its company brochure. The use of the darker gray square behind the logo creates a shadow effect and makes the logo seem like it is hovering above the light gray and white grid that makes up the background.

Direct Mail

This gatefold piece makes use of the same floating logo graphic and overall grid background. Even though the format is completely different from the company brochure, it is instantly recognizable as a promotional piece from this firm by the use of its prominent logo.

RVBK

This architectural firm chose to create all of its marketing materials with the first initials of the partners' surnames. The firm uses both the full name and the acronym. This acronym appears in a loose, freehand typestyle. On some pieces it appears on a gray field; on others it appears on a black field. It also appears in the firm's signage for its office. The firm's materials include letterhead stationery (including envelopes), business cards, mailing lables, company forms, memo cards, special announcement cards, and brochures.

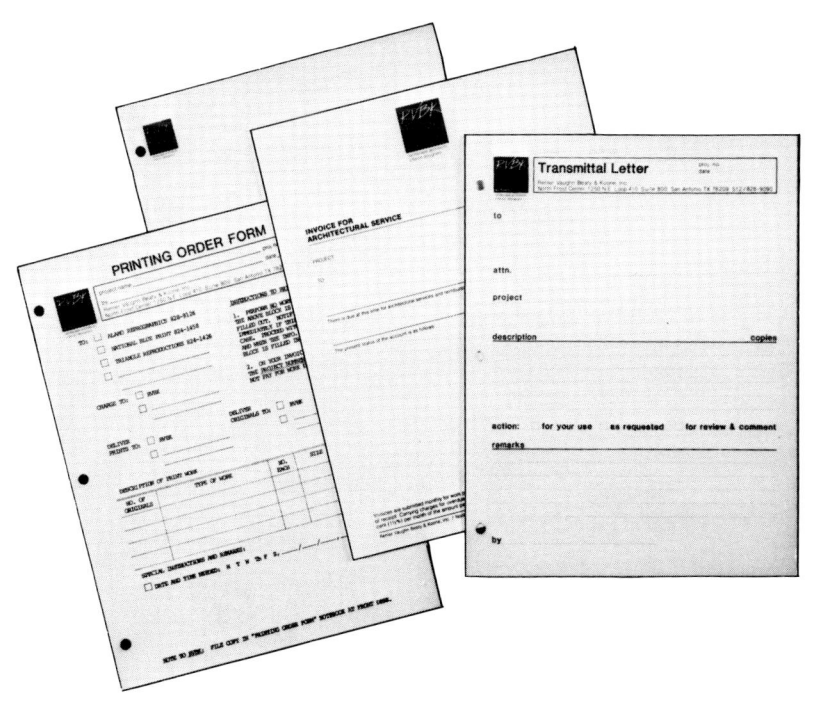

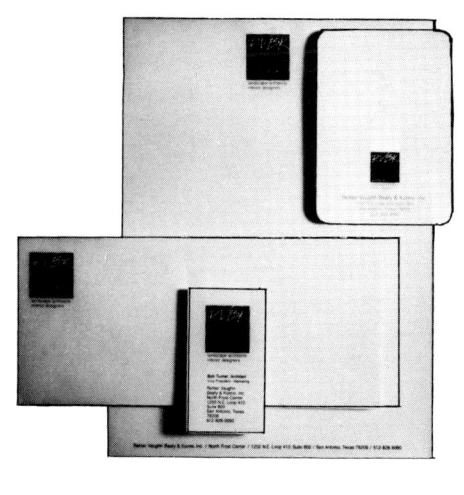

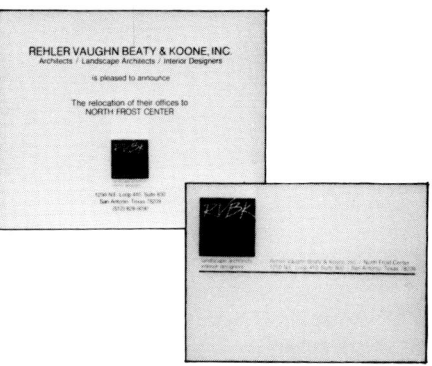

Communications Programs 17

DIVERSITY

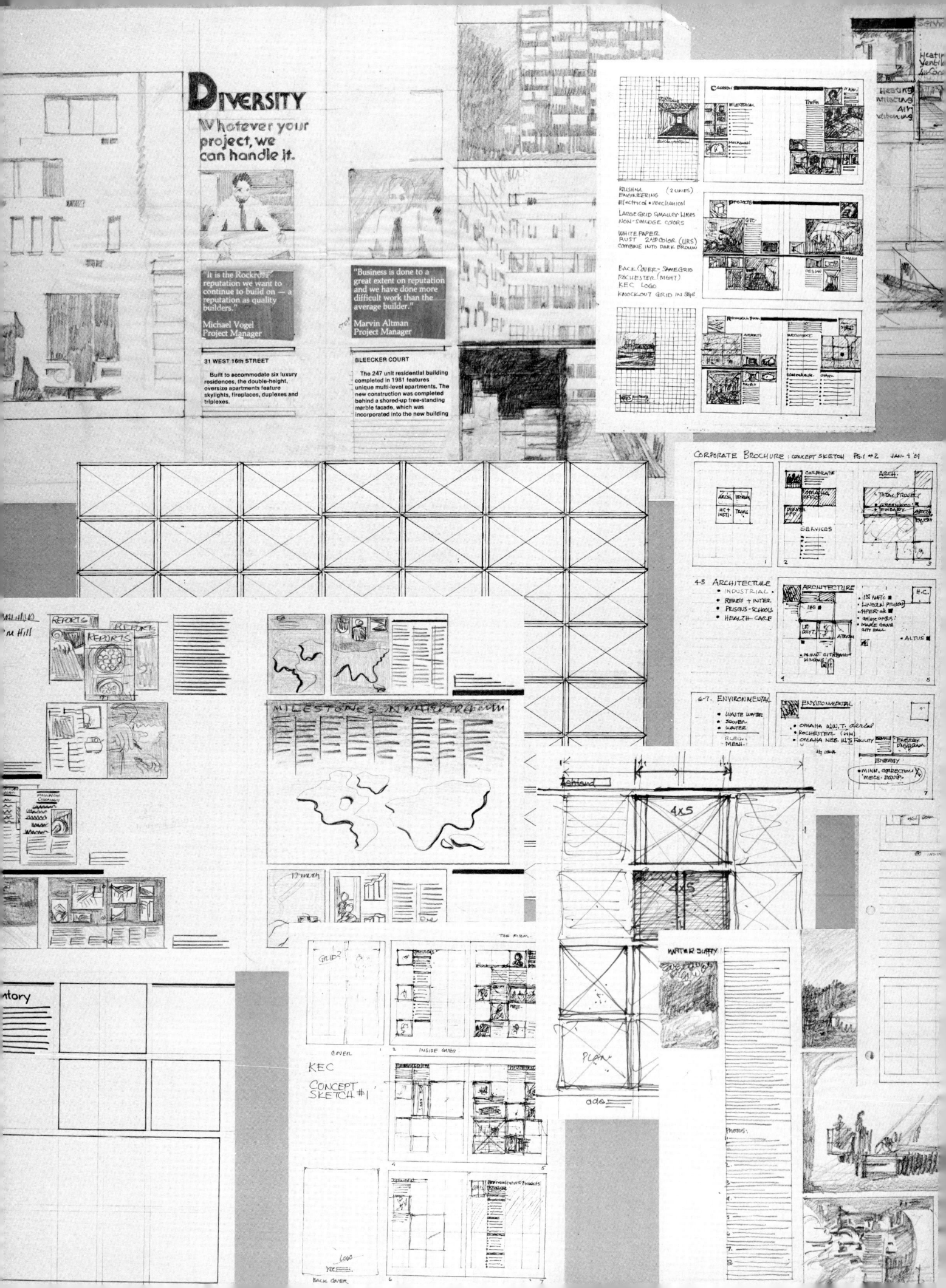

Whatever your project, we can handle it.

"It is the Rockrose reputation we want to continue to build on — a reputation as quality builders."

Michael Vogel
Project Manager

"Business is done to a great extent on reputation and we have done more difficult work than the average builder."

Marvin Altman
Project Manager

31 WEST 16th STREET
Built to accommodate six luxury residences, the double-height, oversize apartments feature skylights, fireplaces, duplexes and triplexes.

BLEECKER COURT
The 247 unit residential building completed in 1981 features unique multi-level apartments. The new construction was completed behind a shored-up free-standing marble facade, which was incorporated into the new building

DESIGNING THE FORMAT: 2

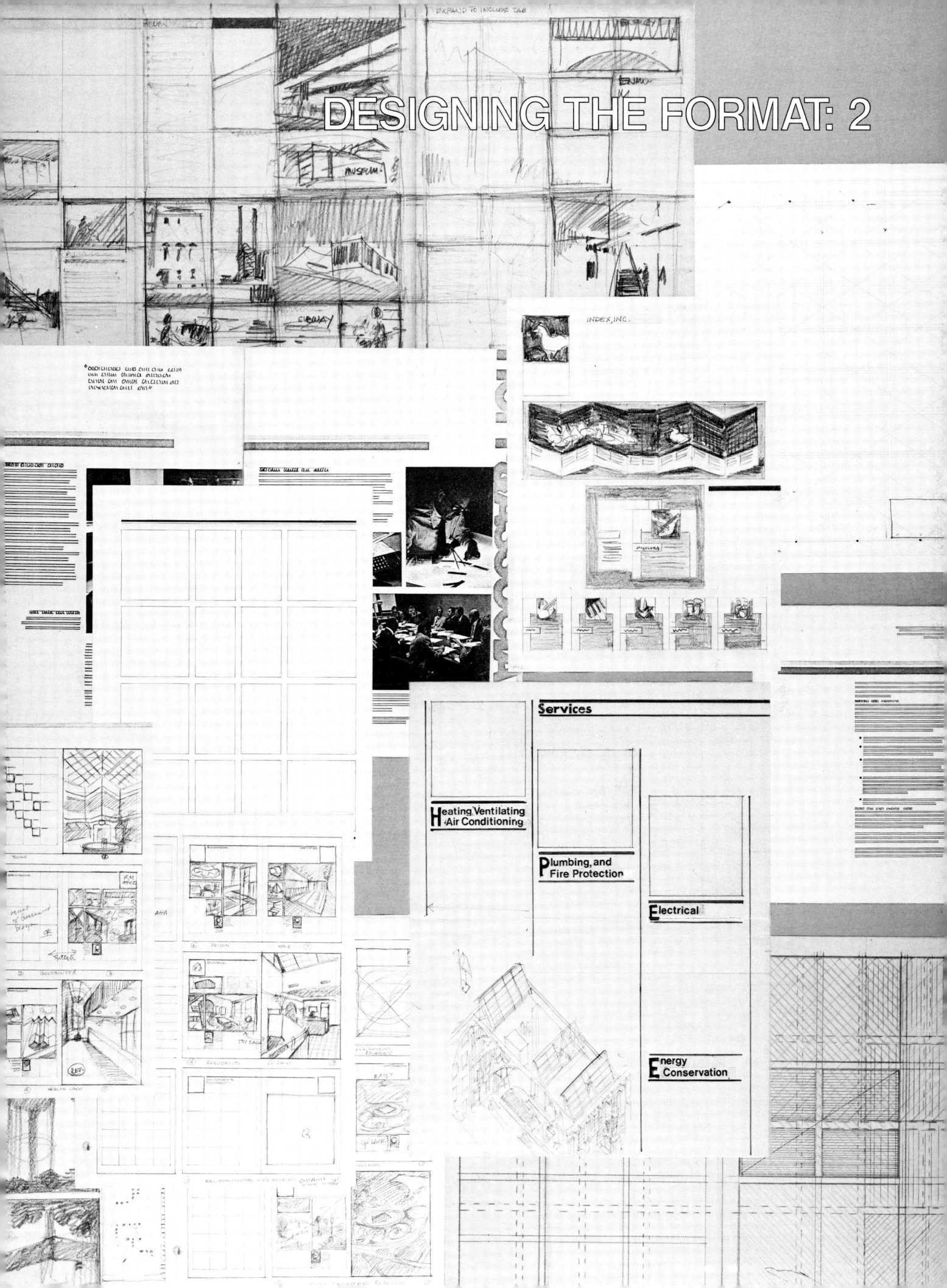

Grid Formats

There's nothing new about the grid. The Egyptians drew grids on stone to enlarge drawings and organize spatial relationships. Graphic designers have been using grids ever since the printed page was divided into more than one column of type. A grid is the placement of one column next to another throughout a publication. It is an elementary concept, but it can be elaborated in complex arrangements, particularly with overlapping grid formats. It helps to standardize the elements of type and photographs, and contributes to the regularity and harmony of page layouts. This is what gives a publication unity The grid for this book was developed around a very specific element, the standard 8½" x 11" inch vertical format that is used for most brochures and the 11" x 17" inch open double spread. Each double spread must be regarded as a single visual entity. In addition to the double spread format the page has to accommodate several brochures, or spreads, on one page so a vertical format of eight columns high was adjusted to a four column wide format. In order to keep the design flexible enough to allow several sizes of double spreads the grid had to work diagonally as well. This meant that any brochure could be shown any size in any grid space, or spaces, throughout the book, and still be consistently related to the open double spread grid element. This type of grid is called a "field" since it works in all directions.

One Column Grid

This grid is useful when a lot of white space is desired. Although it lacks variety, it has a built-in simplicity. There are many options for the placement of the grid within the margins of the page. The only drawback is that it may result in a wide column of text which becomes more difficult to read.

Two Column Grid

This is a traditional simple grid which can be further subdivided for the placement of pictures within the text. Without such subdivisions the photographs are necessarily large and without variation in size. The wide column of type is considered normal for ease of reading, although narrower columns are easiest.

Three Column Grid

This is the most common page layout. It is used for many magazines, newsletters and brochure formats. The narrow column of type is easy to read and the flexibility of picture size is increased. Most newsletters are designed in this format since it is one of the most compatable combinations of type and photograph.

Four Column Grid

This is a distinctive grid, and a variation on the two column format. The columns of type are wide enough to use standard type and can be read or scanned easily. The narrow column permits a lot of variety in picture size and therefore, can create a more interesting interplay of photos and type.

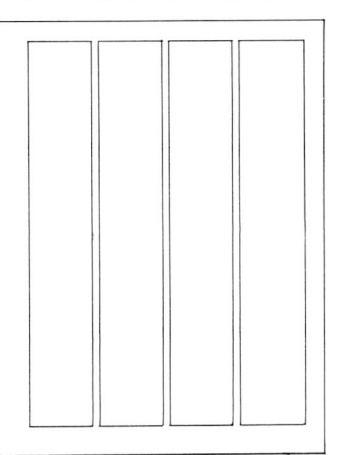

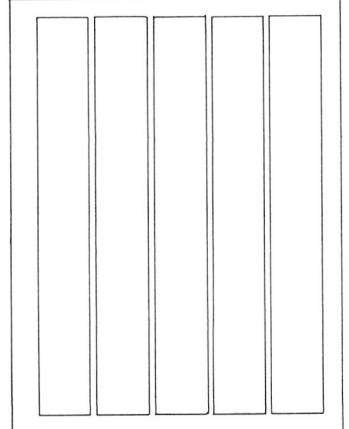

Five Column Grid

The column width is very narrow in this grid and can only be used with small type or ragged-right setting. This will insure normal, or at least acceptable spacing between characters and words. There are many opportunities to combine two of the narrow columns into one for either type or photos giving this grid many more possibilities.

Six Column Grid

This grid is an extension of the three column grid. It is further subdivided for the placement of smaller pictures. The small column may be good for captions but is not well suited for text. There are many interesting arrangements and combinations that are possible with this grid.

Overlapping Grid

There are an infinite number of combinations possible when you use two grids overlapping one another. The most common is the two column and three column grid wherein the center of the panel is divided into two. The variations are possible with both type and photos.

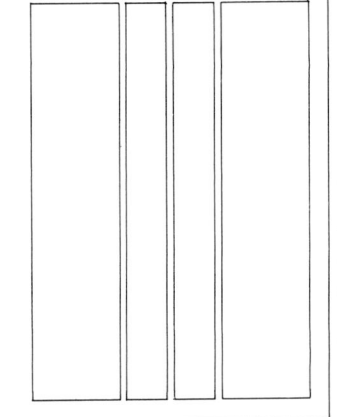

Vertical Grid

The horizontal grid is the most noticeable organizing element for a publication. Additional unity can be achieved by adding a vertical grid which can be based on several options. One such option is the line by line division based on the type point size, and having all the dimensions fall on these increments.

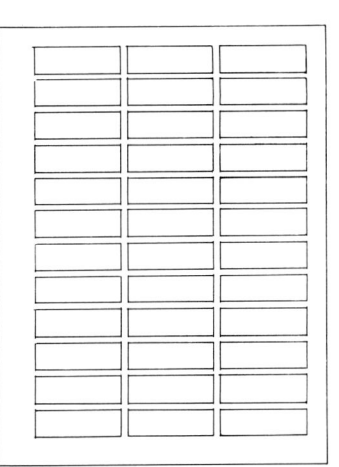

Developing A Sketch Dummy

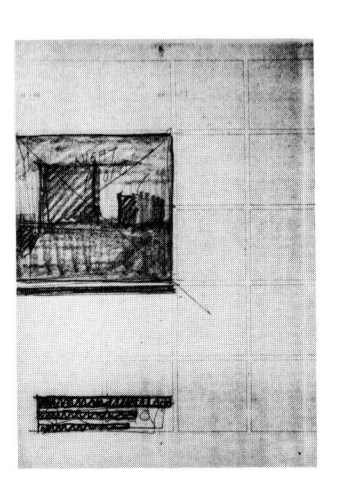

The entire concept of the brochure is evident in the sketch dummy, hence its extreme importance in the production stages. This is where all the research comes together with the goals and objectives of the firm and results in a tangible expression of the firm's direction. There is no other single step in the whole process that is as critical, because the sketch dummy stage is the distillation of all ideas into the physical form of photos and type. If any item is left out of the sketch dummy chances are it will not be in the finished brochure. The value of the sketch dummy is in the flexibility of the medium. The cover and the back cover are important elements and should tell the reader at a glance what the firm is offering either in pictures or words or both.

Next in the order of importance is the services you offer and information about your firm, your staff and all the benefits you can offer to clients. Next in the sequence is some descriptive information about your projects. This can take on many forms, but certainly the importance of a good photographic record cannot be emphasized enough. The next item is your representative clients and these lists are usually found in the back of the brochure although they can also be covered elsewhere. Project location maps are also helpful especially if the firm wants to emphasize geographical service, whether it be a region, state, or the whole world in the case of international firms. The success of the brochure will depend greatly on the thoroghness of the sketch dummy.

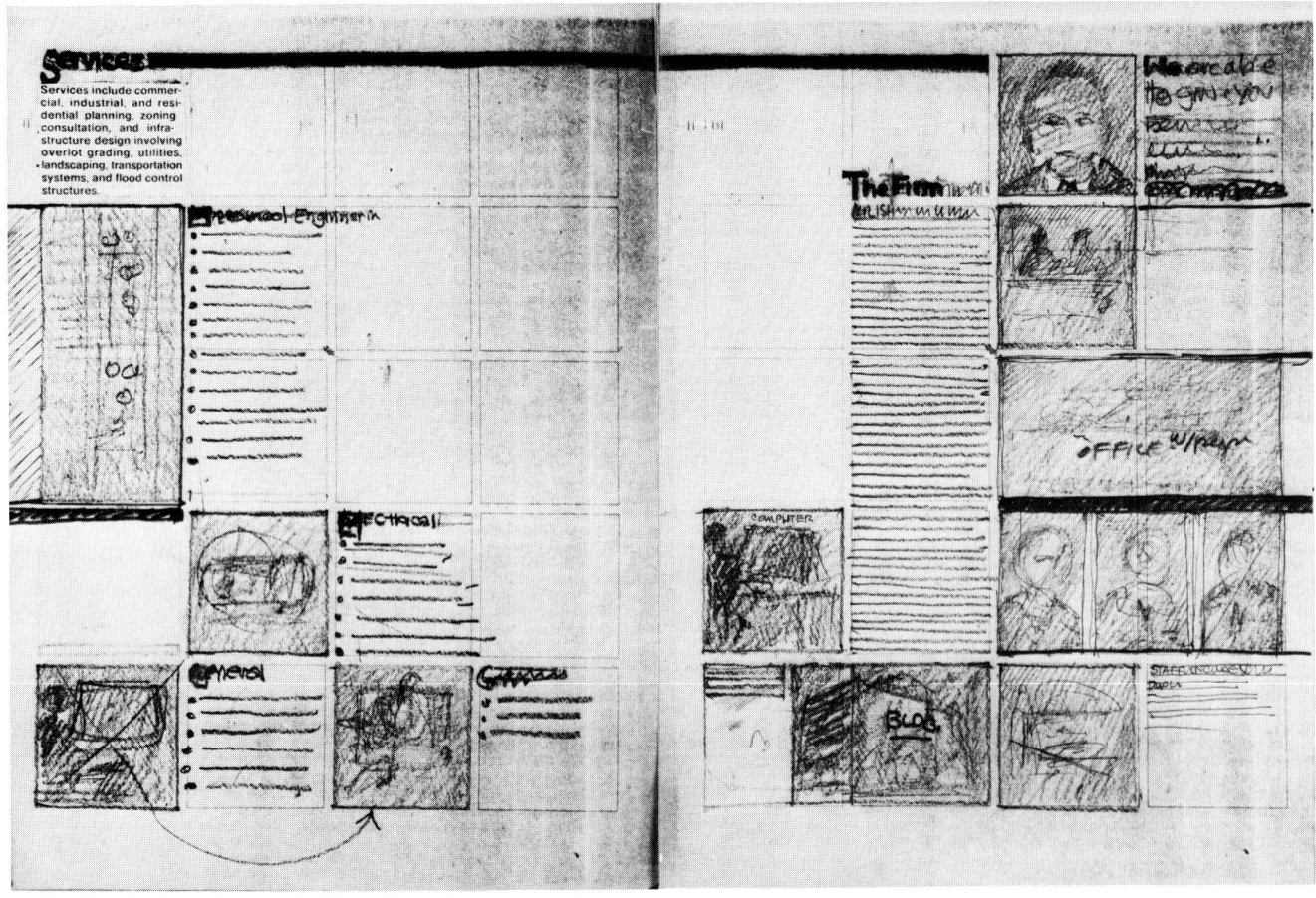

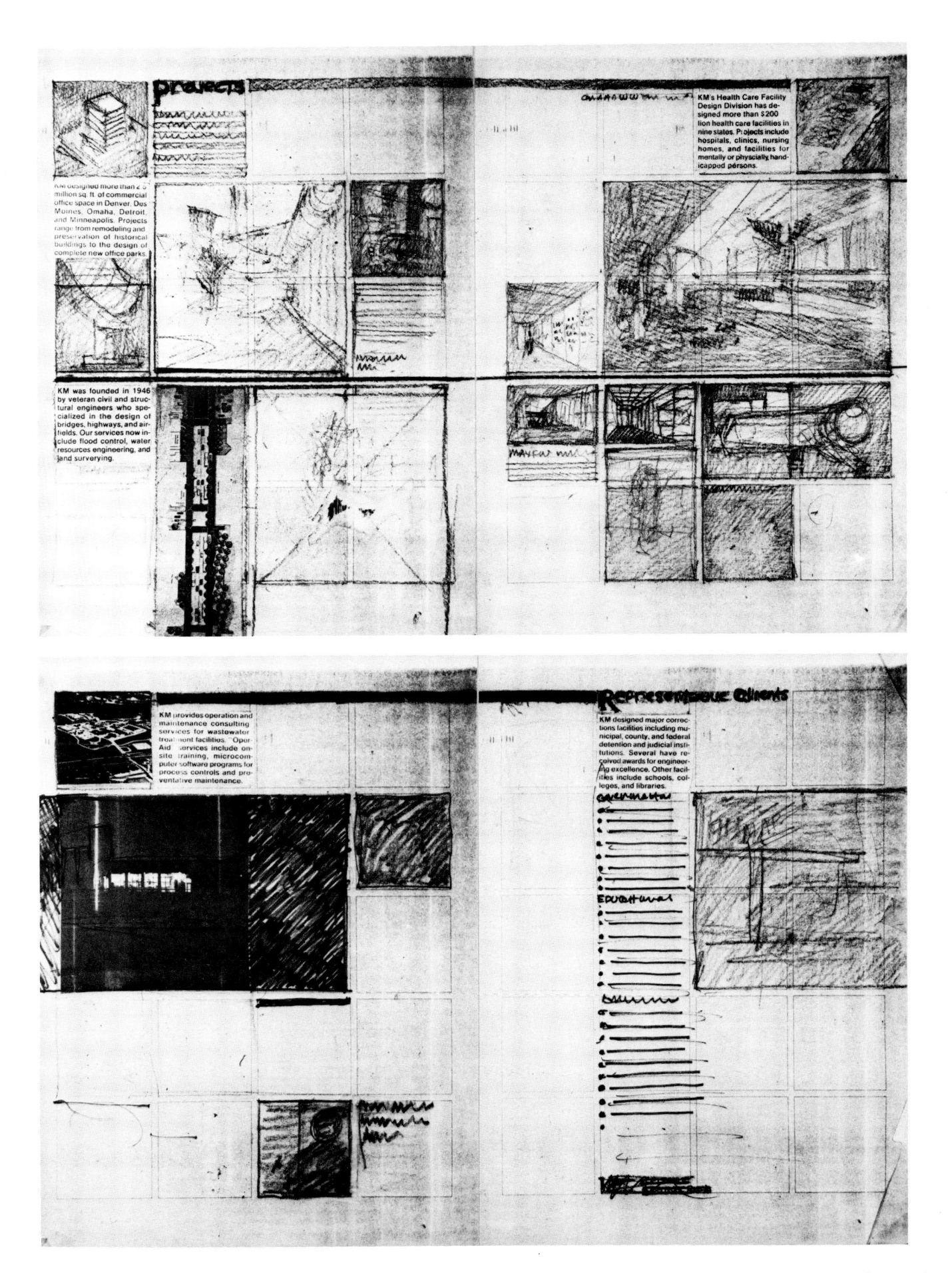

projects

KM designed more than 2.5 million sq. ft. of commercial office space in Denver, Des Moines, Omaha, Detroit, and Minneapolis. Projects range from remodeling and preservation of historical buildings to the design of complete new office parks.

KM was founded in 1946 by veteran civil and structural engineers who specialized in the design of bridges, highways, and airfields. Our services now include flood control, water resources engineering, and land surveying.

KM's Health Care Facility Design Division has designed more than $200 lion health care facilities in nine states. Projects include hospitals, clinics, nursing homes, and facilities for mentally or physically handicapped persons.

KM provides operation and maintenance consulting services for wastewater treatment facilities. "Oper-Aid" services include on-site training, microcomputer software programs for process controls and preventative maintenance.

Representative Clients

KM designed major corrections facilities including municipal, county, and federal detention and judicial institutions. Several have received awards for engineering excellence. Other facilities include schools, colleges, and libraries.

Designing The Format 23

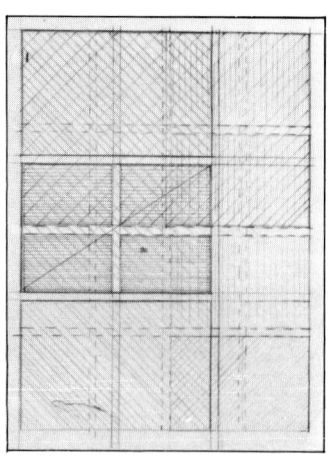

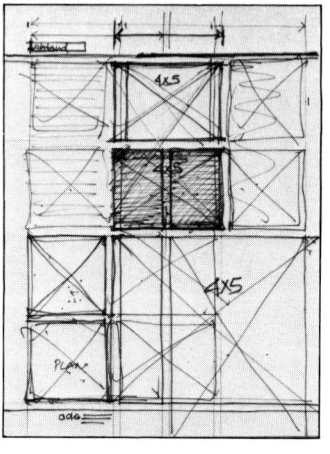

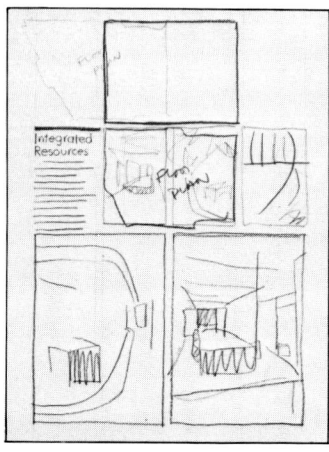

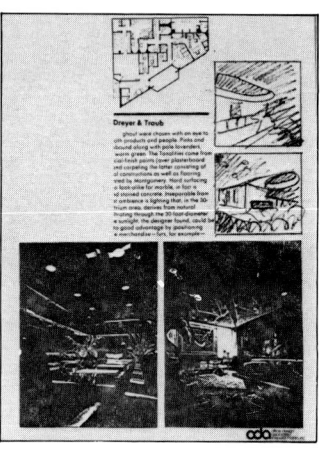

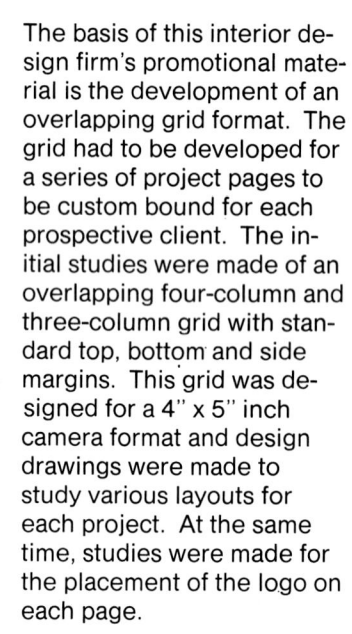

Developing Project Pages

The basis of this interior design firm's promotional material is the development of an overlapping grid format. The grid had to be developed for a series of project pages to be custom bound for each prospective client. The initial studies were made of an overlapping four-column and three-column grid with standard top, bottom and side margins. This grid was designed for a 4" x 5" inch camera format and design drawings were made to study various layouts for each project. At the same time, studies were made for the placement of the logo on each page.

A further development occured when it was discovered that in addition to the existing 4" x 5" inch images, there were also many projects photographed in a 2¼" inch square format, typical of many interior projects. Thus the grid was further refined to encompass both formats in an overlapping two-and three-column format. Each project was then resketched on the basis of the newly adopted grid format. The placement of the logo and the grid module's ruled lines would now appear in the exact same location on every page. More exact photo dummies were then produced of each project.

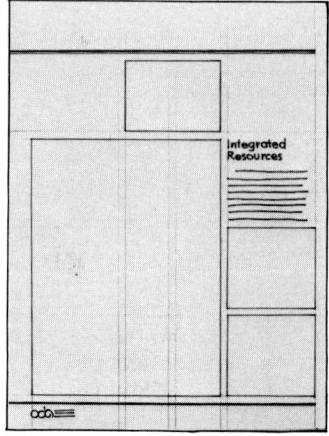

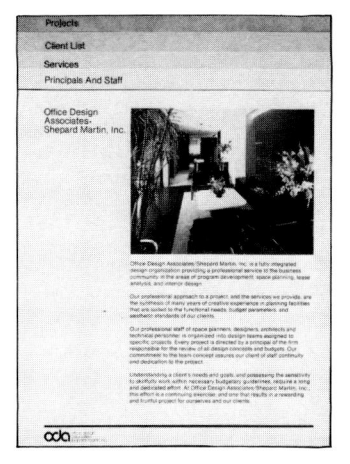

The firm's promotional package was designed to contain the following elements: an introductory statement on the firm, resumes of the principals, client lists, project pages, magazine article reprints and a blank grid module page for customized messages. The introduction, resumes, client list and project introduction were designed to stack in the booklet with the first page trimmed to expose the other four and the second to expose the other two, and so on.

The firm wanted to coordinate all its printed material, so previously published articles were redesigned into the same grid format. The original article was reset in type to match the other pages and the photos were resized for the new layout. The names of each publication and project were printed on the front cover to make it look exactly like a magazine article reprint. The balance of the promotional material was provided by individual project pages. Some featured reduced scale floor plans of the space and a very brief description set in the type style that was used throughout the entire package. The overlapping grid module allows a great deal of flexibility in the layout and arrangement of photos, making each project page unique-looking within the unified package.

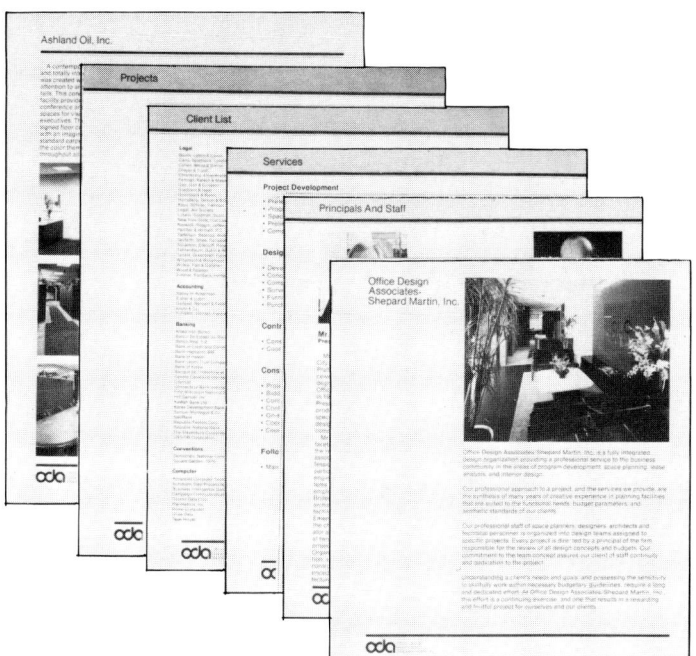

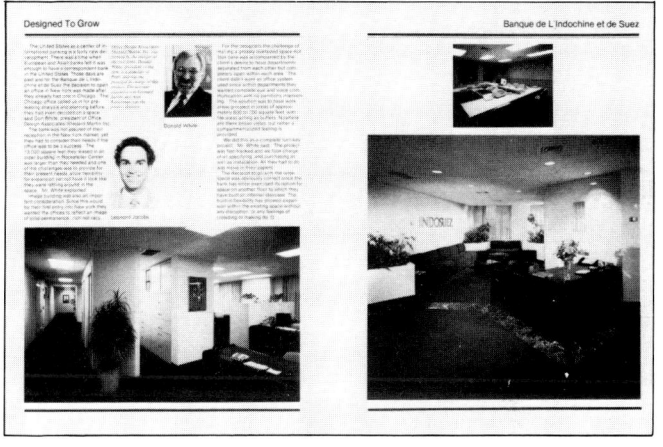

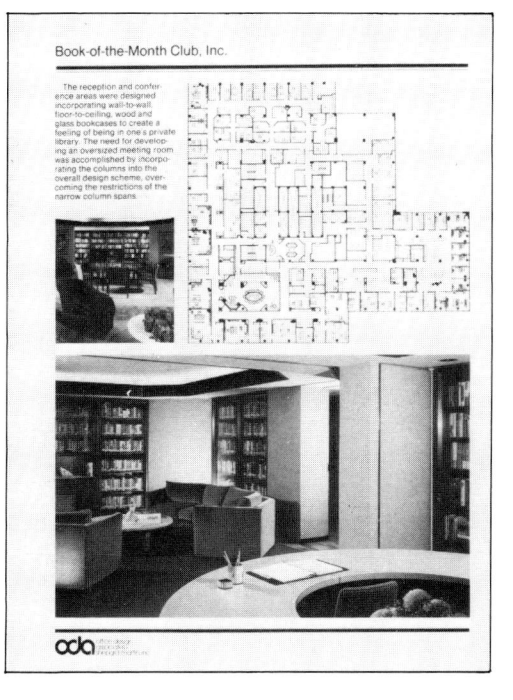

Designing The Format 25

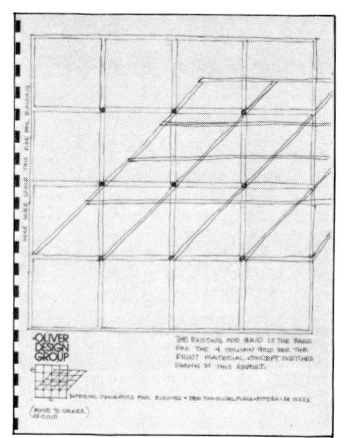

Developing A Grid Format

This interior design firm had already developed a corporate identity program for its logo and stationery. The strong logo design became the basis for the grid, which controlled the development of additional collateral. The logo was enhanced to full-page size and the four-by-four square grid design was developed. The firm's logo was positioned at the bottom of the page on all pieces. Sketches were made of the various pieces in the promotional package such as client lists, project data sheets and color project pages. The visual material is confined to the square area within the grid.

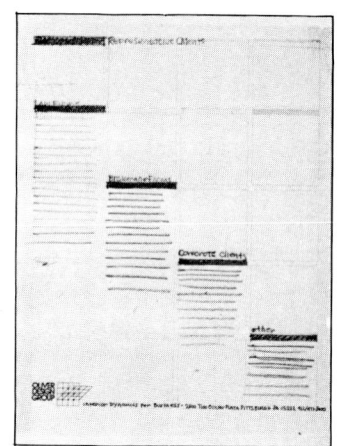

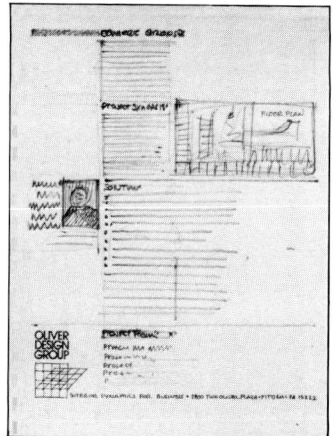

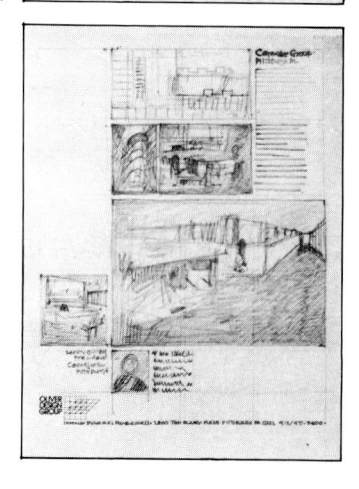

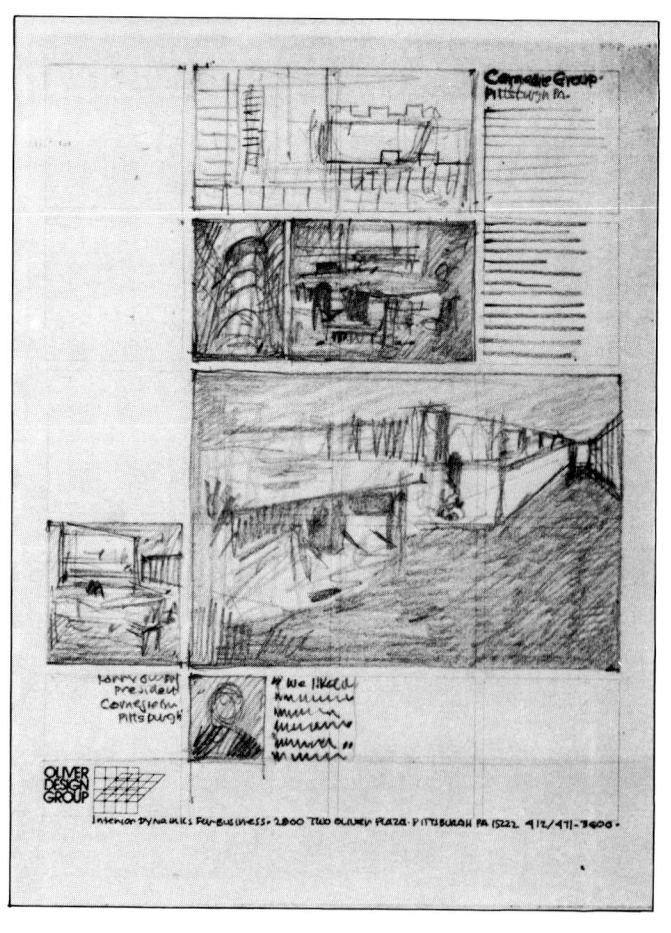

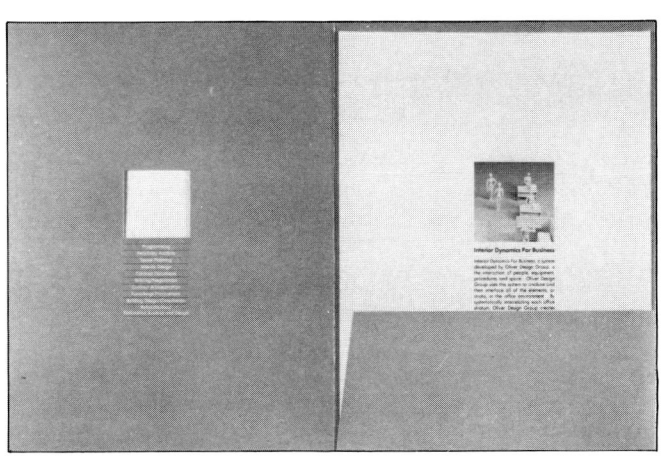

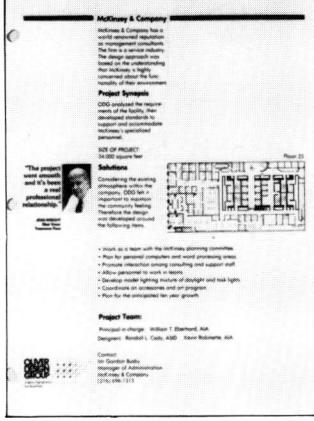

A folder was designed to hold various promotional pieces. It has a die-cut window to expose the sheets inside. The package consists of a list of representative clients, one for representative projects, and a detailed sheet called a project synopsis. The main color project pages feature several views and a photograph and quote from the client of the project. This flexible series of color pages could easily be customized for a particular prospective client or spiral-bound into a proposal.

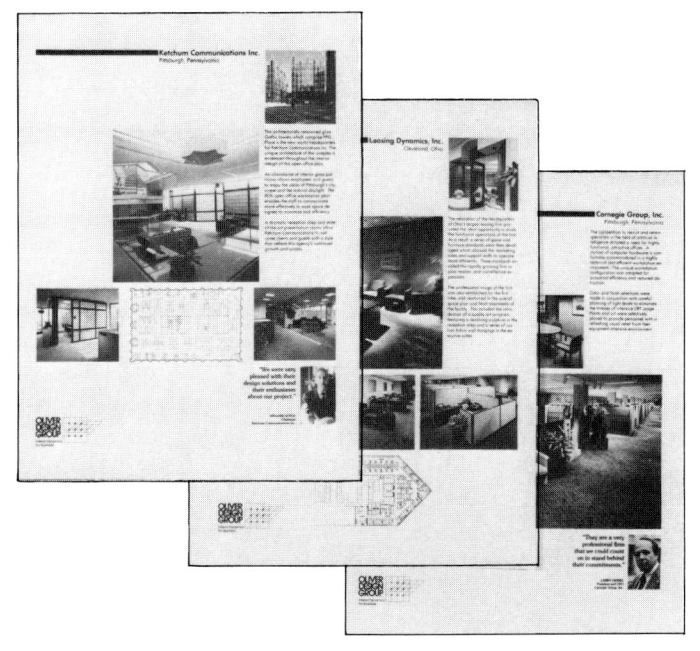

Designing The Format 27

Developing A New Format

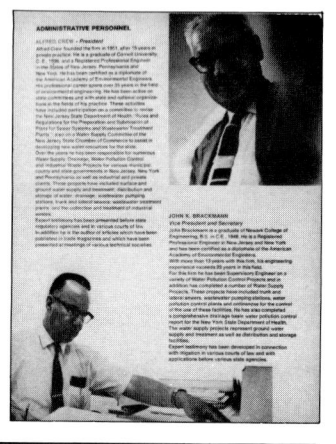

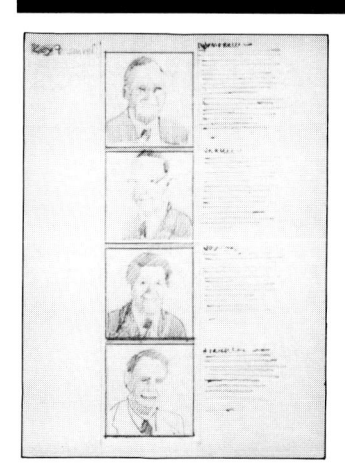

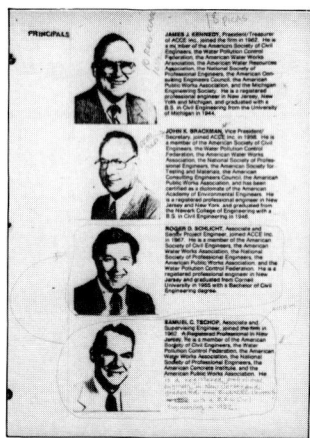

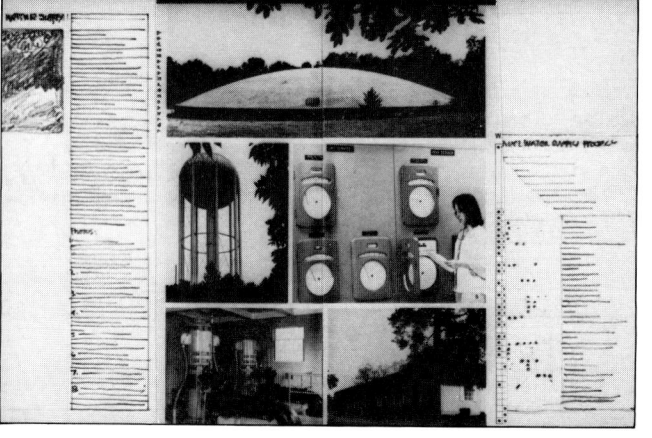

Most firms who have been in business for more than just a few years have already produced some form of a company brochure. At the very least the firm will have accumulated a lot of photographs documenting its projects. Sometimes these photos are simply snapshot-quality prints taken with an inexpensive camera.

In the case of this environmental consulting engineering firm it had an existing twelve-page brochure plus a four-page cover. The stock used was a heavy textured and lightly tinted gray paper. Further, there were many photographs that were printed with a color-screened tone behind the picture. This gave the brochure a drab and uninviting look. The quality of the original photographs however, were very good, so it was decided to re-use existing material

Design of the interior pages began by taking formal portraits of the principals and featuring them in the beginning of the brochure.

Each double spread was then treated as a single unit, and featured one of the firm's services. First, a rough sketch was drawn up using the existing photos as a guide. A theme photo appeared in the upper left-hand corner of each spread. The type next to this theme picture was designed to be set in semi-bold type to capture the attention of the reader. Below this introductory text was an identification and brief description of each project illustrated. The project photos were contained within a square format which included photos that went across the gutter of the double spread. The right-hand column of each spread contained a custom designed matrix chart which listed clients, projects, and services pertaining to each of the firm's services. The organization of the statistical information combined with the visual impact of the photographs contributed greatly to the clean and simple appearance.

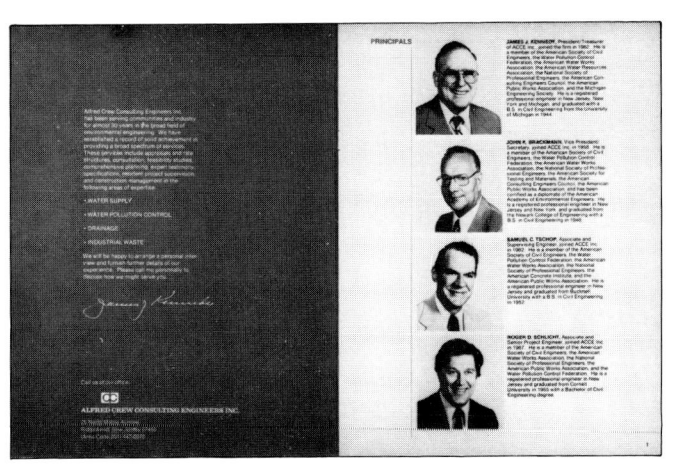

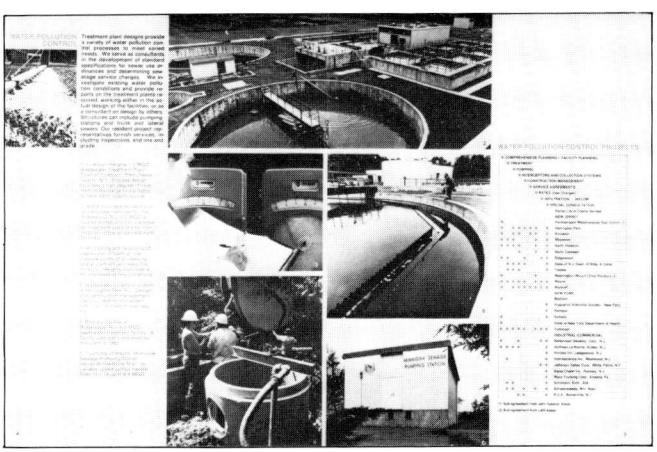

To give the brochure a fresh new look, a bright-white coated paper was selected for the eight-page interior and a heavier weight high-gloss coated stock was selected for the cover. To add a totally new look, the cover was designed as a full-page bleed color photograph of one of the firm's bridge projects. The low angle view was selected from a series of 3" x 5" inch color snapshots. This view placed emphasis on the stream rather than on the bridge and culvert. The name of the firm and its specialty is dropped-out in white type.

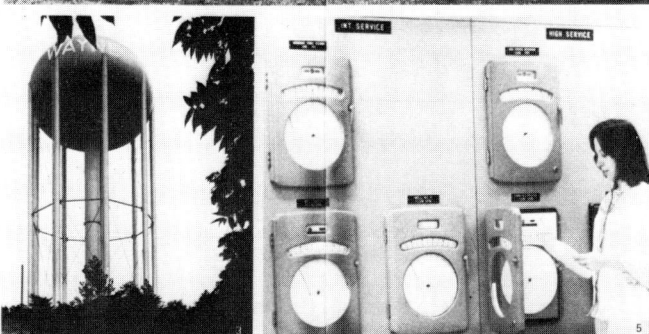

Designing The Format 29

Developing An Expanding Brochure

Most consulting firms want the most flexible brochure system possible, one that can be customized for each project presentation. The services provided are nearly always the same so a method of segmenting those services has to be devised. One method is to develop a series of folders featuring each service or project type. Initial designs for this mechanical and electrical engineering firm's brochure began with studies of three-panel foldout fliers. The studies were based on a four-column grid layout which represents the four services that the firm offers. The studies showed that the projects could not be separated properly and that general confusion would result from that kind of layout. In addition, it could not be expanded by project type so a new layout was studied. This new layout was based on a three-column grid which related to the three principals of the firm. Studies of lettering were made on a large scale and reduced for the final design layout. The flexible system that resulted consisted of a four-page cover, a four-page colored paper wraparound that was inserted between the cover and the four-page inside. This makes a substantial bound brochure for general purposes.

When bound for proposals, the brochure is trimmed off at the folded edge and individual project sheets are added to the package. They all follow the same grid format as the main brochure and expand the information through selected project coverage. Since the technical information and data on each project are important to prospective clients, one of the columns is reserved for this information. The format for this information is the same from project to project. Photos and drawings of mechanical and electrical installations were chosen for some of the illustrations rather than simply showing a completed building, which would not have been representative of the firm's work. The total spiral-bound package has a consistant, easy-to-follow and thorough description of the firm's specialized capabilities.

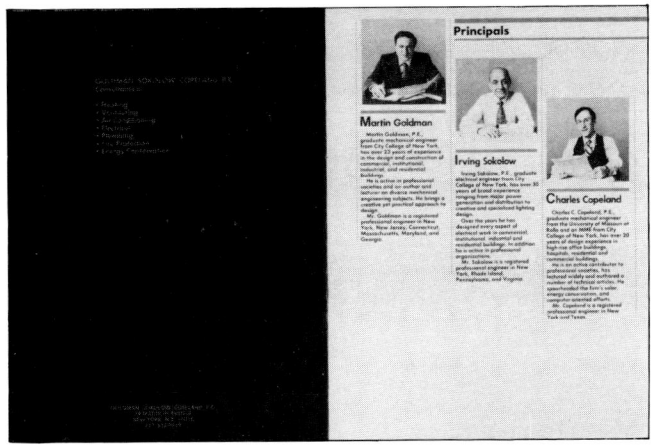

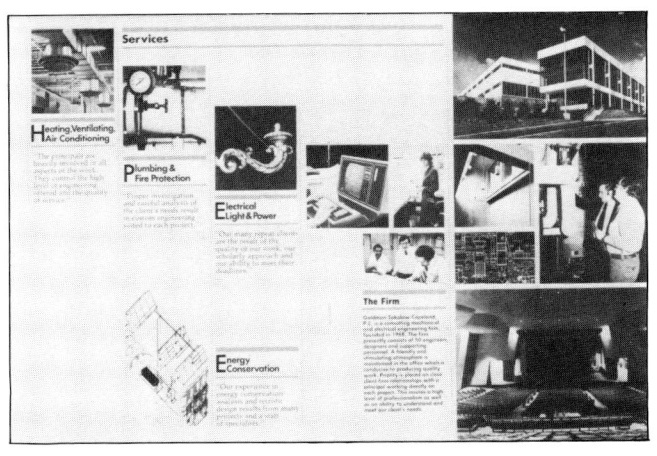

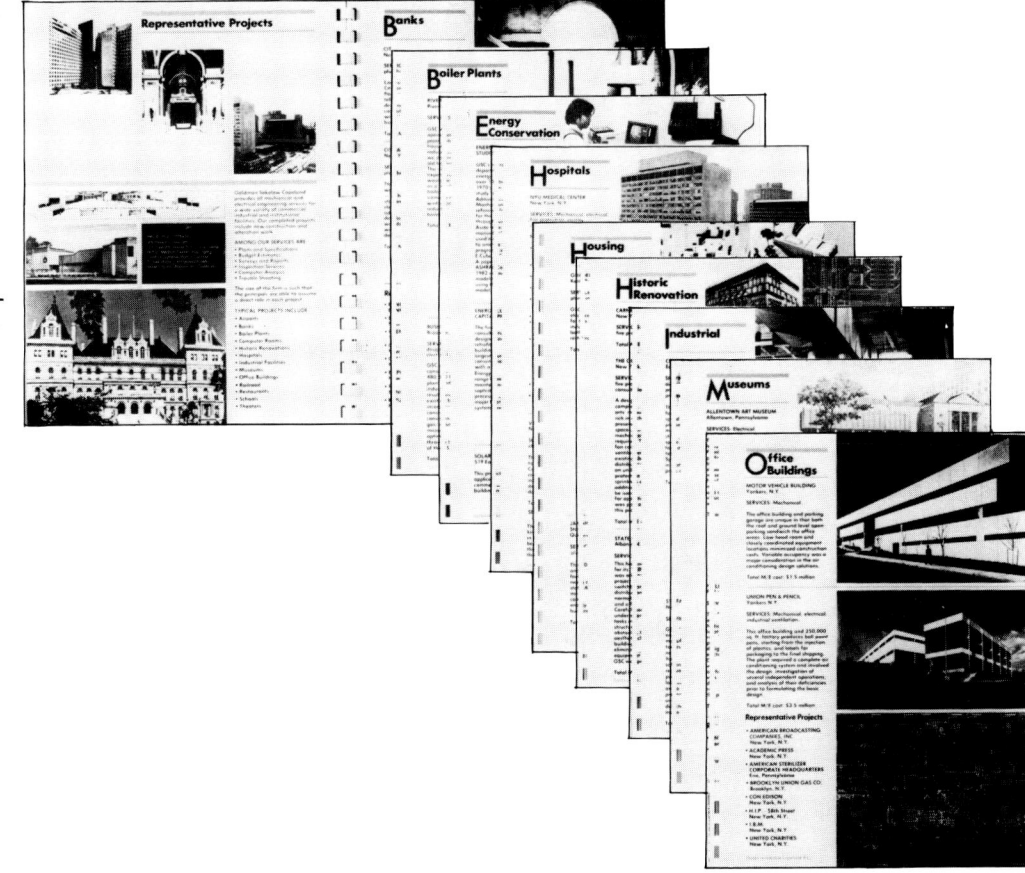

Designing The Format 31

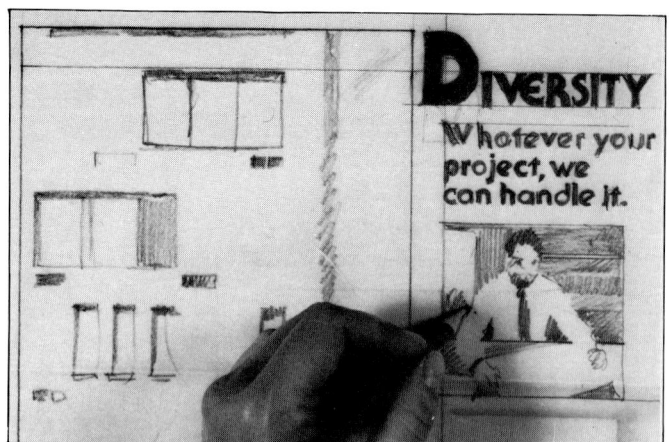

Developing A Company Brochure

The primary purpose of a brochure is to attract the kind of work you want to get as opposed to simply showing what you've done in the past. This is true of this contracting company's first brochure, which was designed to get construction management projects.

The design of the brochure began by taking inventory of all existing photography. It was clear from the collection of photographs that most of the firm's work from the past was renovations rather than new construction. Therefore the design would have to overcome the image that the firm was limited to renovation work. Several new photographs were taken of key individuals showing management-related activities rather than on-the-job construction shots.

Following approval of this sketched dummy, rough color prints were made to study the layout more carefully. In this photo-dummy stage it is not necessary to have every component in its final form, because many changes may be made at this point. Therefore, color photocopies, polaroids, and dummy type from other brochures may be quite useful in achieving the desired affect. Then it's an easy matter to create only those items you are missing.

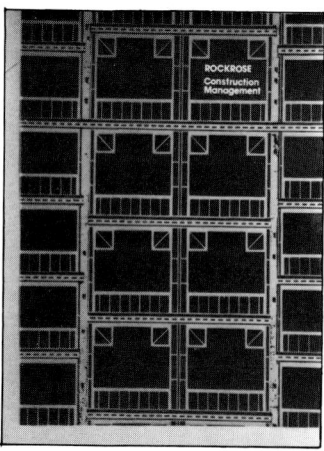

Design studies of the cover are also important. In this case the initial choice was a carefully composed photograph of a newly completed project. An alternate design was also considered, using an abstract pattern from several floors of a building under construction. It was decided that both approaches suggested construction rather than construction managemant. The final choice was the use of the words "ROCKROSE: Construction Managers" and the firm's logo in a black field. Inside, this theme is carefully carried out

32

The emphasis was placed on the level of expertise of the firm's key people, on the inside front cover and on the inside back cover, since some people open the brochure from the back. The main spread shows a large project underway and the activities illustrated relate to construction management rather than construction.

To give the brochure a sophisticated look and to further emphasize the carefully cropped photos of the buildings, the background of each page is designed with a matt-black field. The photos were treated with a high-gloss varnish to make them pop out of the dull black background.

By combining carefully cropped photographs of all the firm's work to make them seem like new projects, and by the use of bold headlines, the desired image of this company was achieved.

Additionally, the company wanted to supplement its reputation for award winning renovations and new construction projects. Therefore, two supplemental folder were designed, one for new projects and one for renovations. These folders follow a similar format to the main brochure but are printed in two colors instead of full color.

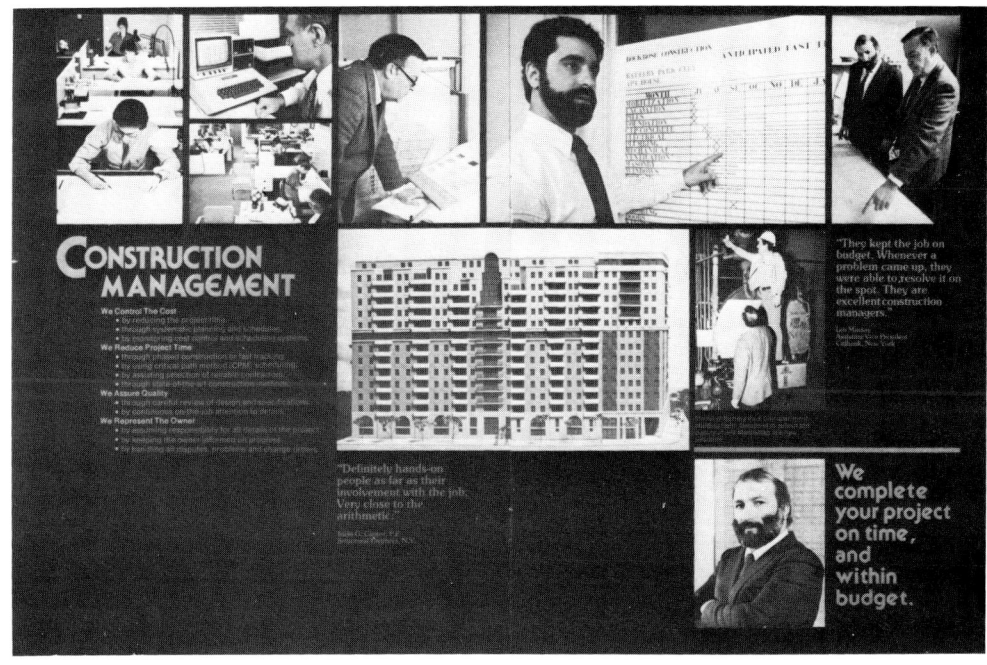

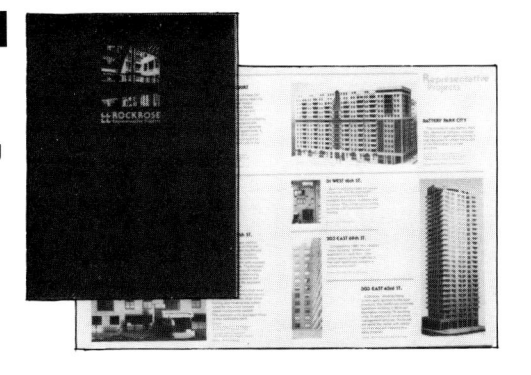

Designing The Format 33

Developing Corporate Materials

In a large multi-disciplined firm the coordination of promotional material becomes more critical. It is further complicated by branch offices or divisions within a firm. A corporate promotional materials master plan diagram is a good way to plan the relationship of the pieces to each other. The diagram is similar to organizational charts where a hierarchical relationship can be established.

The corporate brochure is a company's primary marketing tool. One off-shoot of the corporate piece is a vest pocket brochure, or promotional flier, which can be sent out to solicit inquiries. Based on the response, other pieces are sent.

Within a system of material, there are many possible formats that can be used either singularly or in a combination. Pocket folders are the most common front line item for holding materials. The folders can be printed with information to enhance their utilitarian value.

A diagram of how each piece is coordinated with the total package is helpful in visualizing the relationships of each piece to the various disciplines or branch offices. These three diagrams form the basis of a corporate-wide promotional material master plan. Other support pieces to the master plan are the newsletter, project folders, resumes and photo project pages.

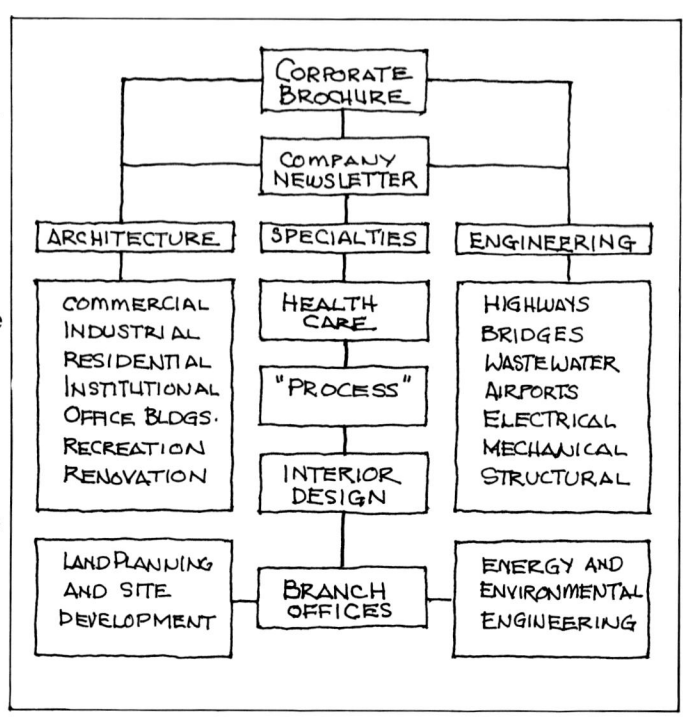

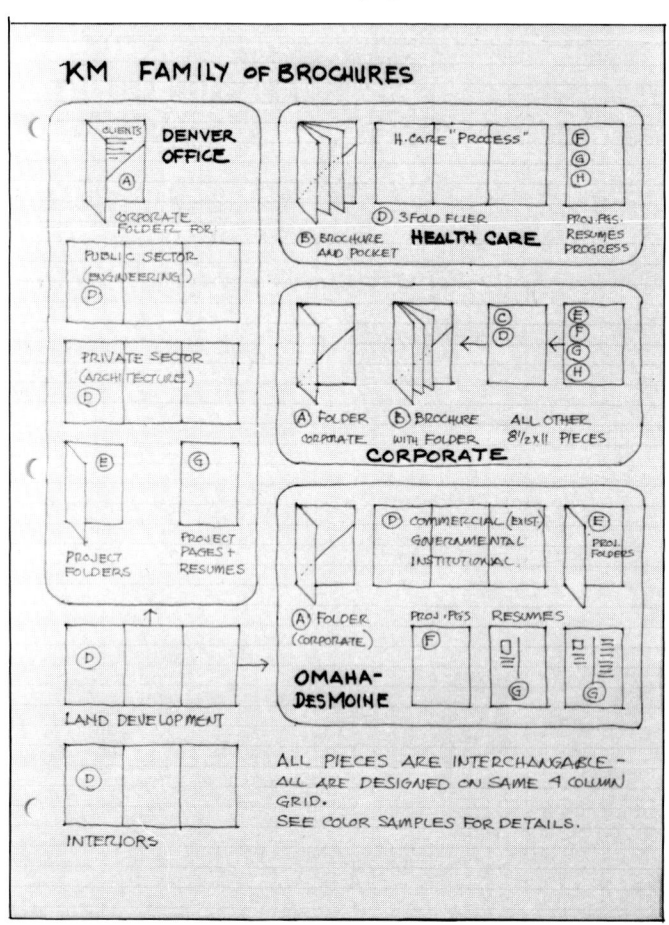

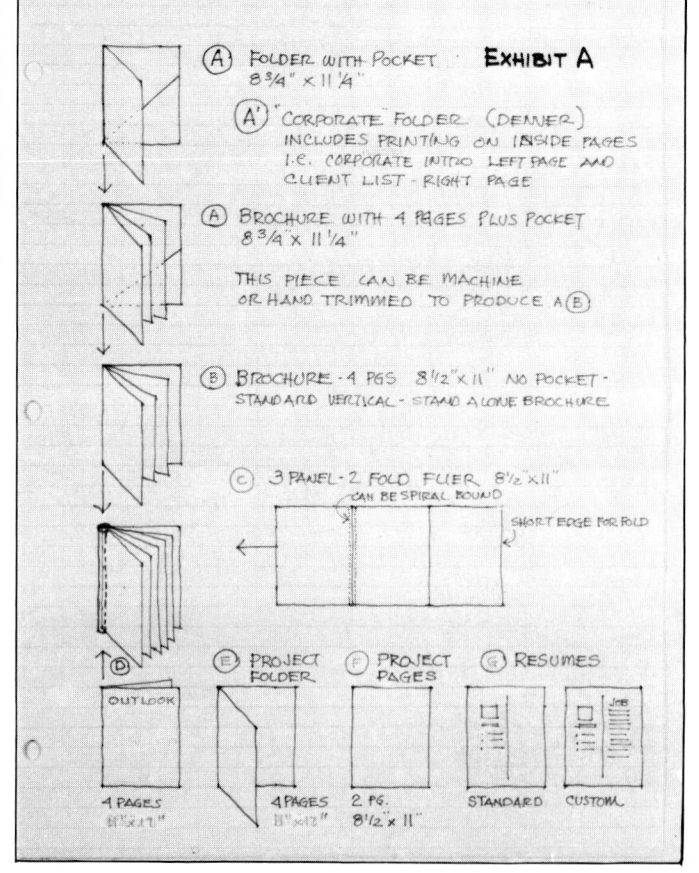

Once a design grid is developed, it should guide the layout of all of these pieces. In this case the grid is a four-column wide by five-column high square grid. The firm's newsletter is layed out on this grid, which allows a lot of flexibility in the arrangement of photos and text. It also features a photograph and quote from the client of the project featured. Four-page folders also follow this four-column square grid.

Color project pages which are designed on a grid, have minimal descriptive text. All the pieces have a liberal amount of photographs. All of this material can be combined and presented through the use of a folder or by spiral binding them for presentations. Next, the last page of a brochure can have a pocket to hold additional material. This slightly oversized format can be trimmed off with a knife for a stand-alone clean piece to go inside another folder. Most other pieces, such as resumes and client lists, should be designed as standard 8½" x 11" inch vertical pieces, or as 11" x 17" inch, four-page folders for special projects. Another special format is the 11" x 25½" inch, three panel, two-fold which is very popular for specialty markets or discipline brochures.

Designing The Format 35

Developing A Corporate Brochure

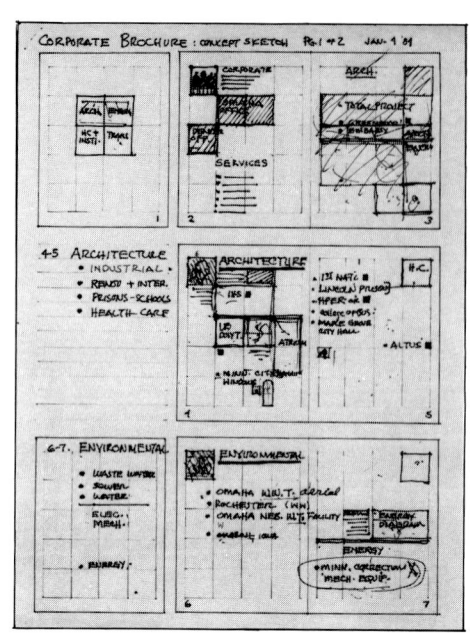

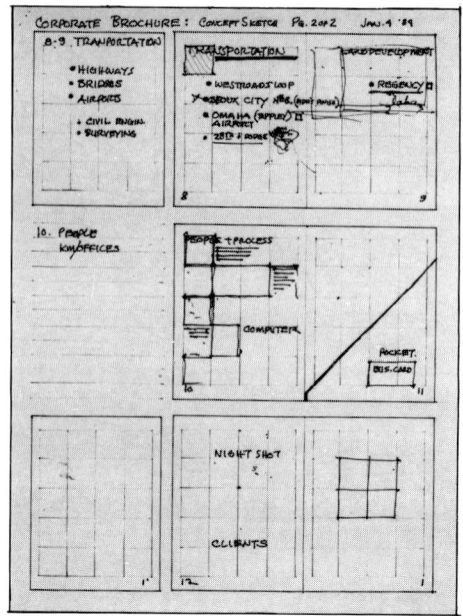

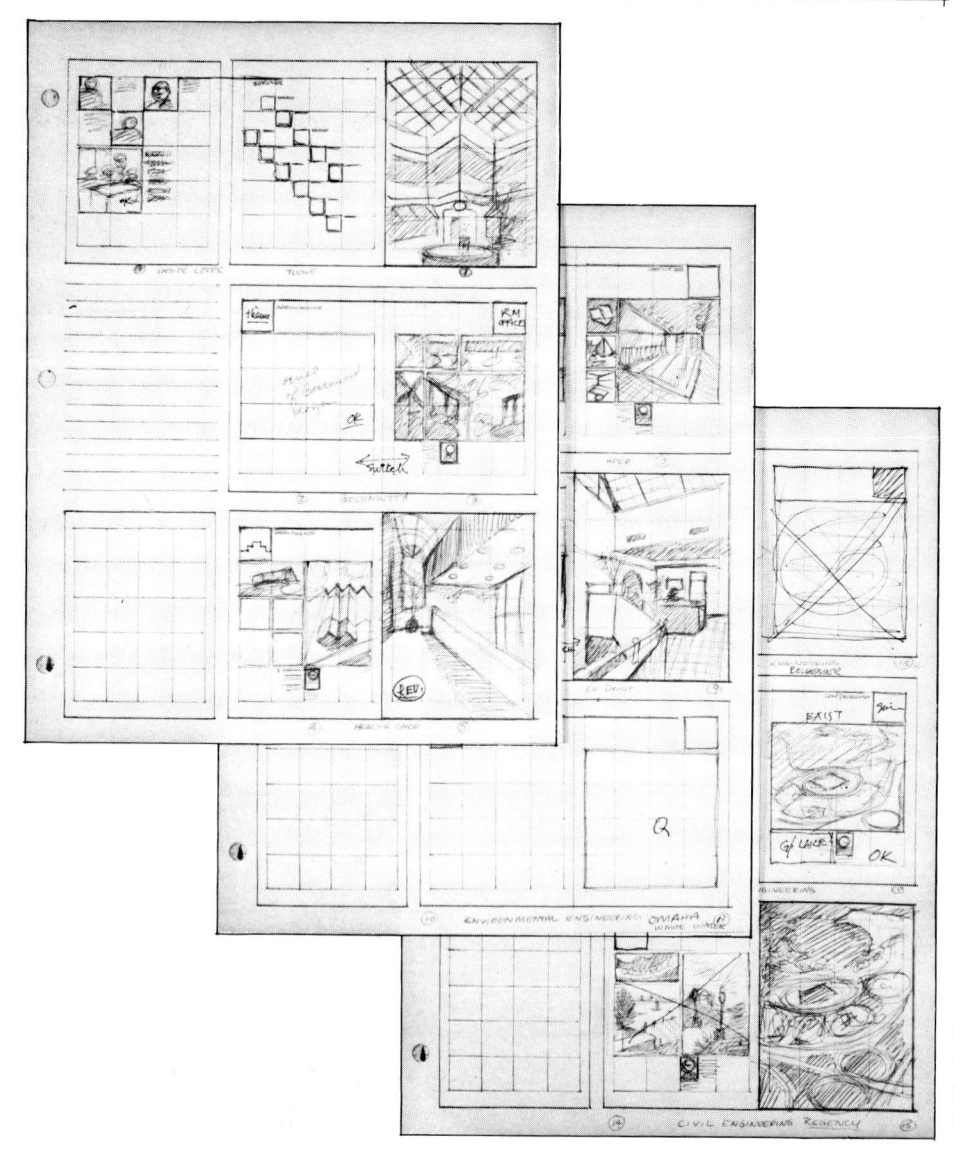

One helpful method of designing multi-page brochures is the thumbnail sketch or miniature layout. This allows you to see all the pages at one glance and to make decisions quickly as to sequence and placement within the brochure. Although the sketches may be too small to aid in the photo layouts, many variations can be studied at this stage of design and many decisions can be made.

The next stage in the design development of the page layouts is a mini-page where three reduced size double spreads can fit on an 8½" X 11" inch standard page with room for a column of notes. These miniature grid pages can quickly be drawn in pencil or pen and photocopied for as many design studies as it takes to get approval for the concept. Then you can begin full-size sketch layouts which will carry the design further. Following the sketch dummy is the photo dummy where photographs or photocopies are used to approximate size and placement to study the page layout. For example, black-and-white and color xeroxs can be used in place of the actual photographs. Enlargements into poloroid prints can be made from 35mm slides to study picture selection. The composition of the final selection of photographs will be done more carefully in the final mechanical preparation stage. There are other time saving techniques such as the use of a hand waxer to position the photos on the dummy so that they can be easily picked up and moved around until the final layout is agreed upon. Also, dummy type from other brochures can be used in place of the final type selection to give the dummy a more finished appearance. If a colored background is going to be used, colored paper should be used on the dummy or comp to study the appearance of the photos and type against it.

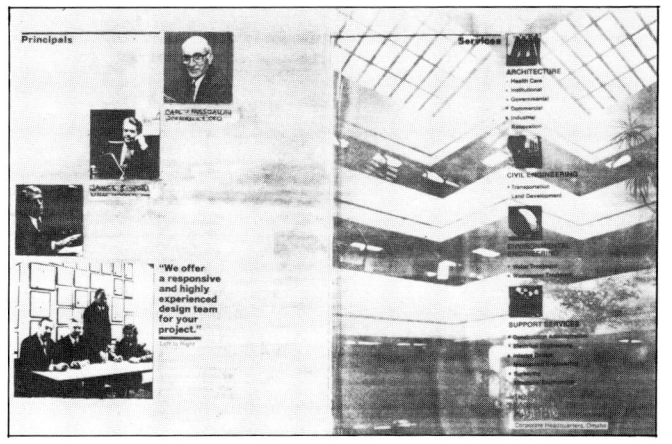

In the case study here, the corporate brochure cover has a square grid with four pictures in it. These appear in the four sections within the brochure indicating the services illustrated on a particular double spread. The opening page shows the key individuals in the firm opposite a tissue overlay with the entire list of services. The column of information on the tissue overlay is designed to line up with the corners of the interior atrium. Each typical spread has one large photograph controlling the visual impact, complemented by smaller photographs on the balance of the page. Photographs of clients identified with a quotation about the firm's services appear throughout the brochure. The inside back cover has a diagonal pocket to hold other company material.

The back cover of the corporate brochure established a theme that was followed throughout all corporate pieces, the use of dramatic nighttime photographs of one of the firm's major projects. The location of all branch offices are dropped out of the black background.

Designing The Format 37

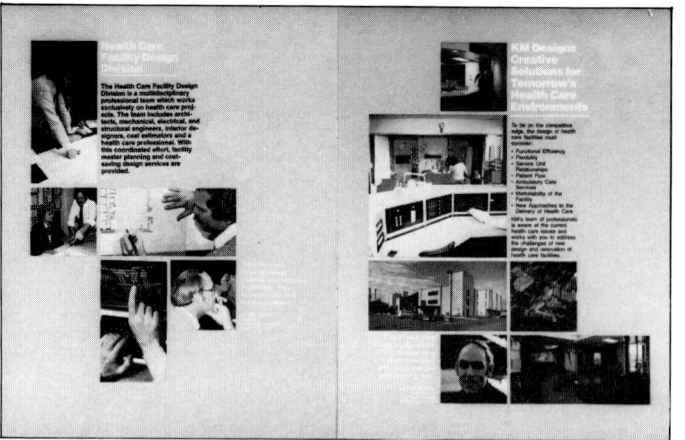

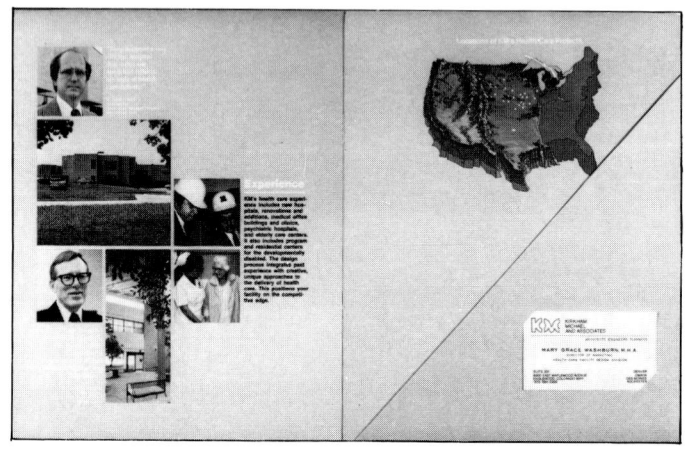

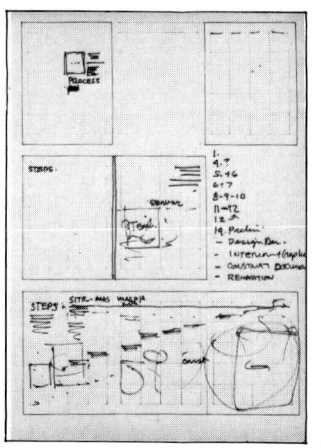

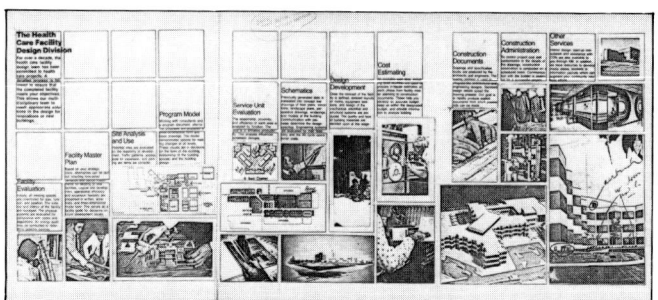

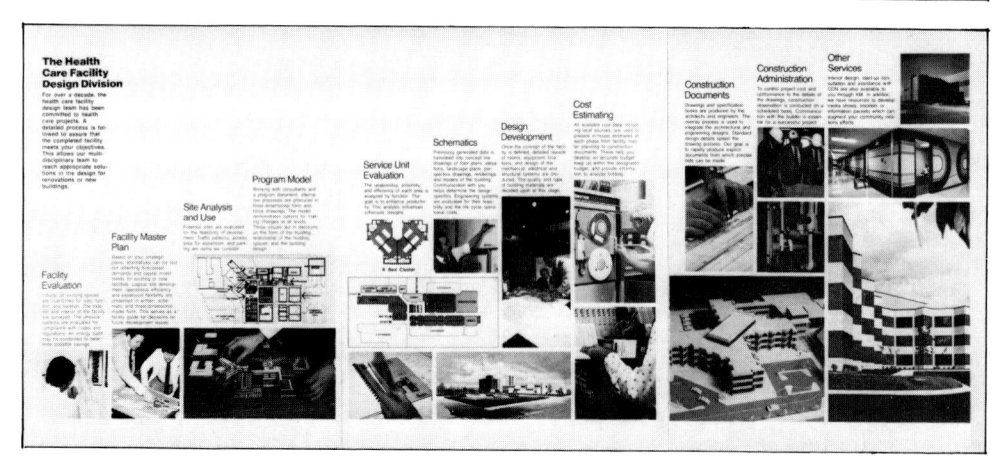

Developing Specialty Brochures

The firm's health care division produced two brochures, one is an eight-page brochure with a pocket and the other a six-page, two-fold flier that fits into the pocket. Each is designed on the same four-column grid which is expressed through a high gloss varnish on the cover and on the inside of the three-panel folder. Each spread of the main brochure features projects and clients, the back page has a map showing client locations.

A popular and efficient format for specialty brochures is the six-page, three-fold folder or flier. It can be printed without binding expenses, and is easy to open and read at a glance. Because it is an unusually long format, (consisting of two folds) it should be printed on heavy paper stock for ease of holding. These brochures are usually designed to supplement other promotional materials. In the case of Kirkham Michael, this format was used to cover four separate markets; health care, commercial architecture, and a public and private sector brochure for one of its branch offices. Each of these pieces could be inserted into the back pocket of the main brochure.

The three-panel specialty folder for the health care brochure features the firm's approach to the health care "process". Concept sketches were made of the cover and inside spread and then were further developed into rough design sketches. Here the idea of a sequential "build" from the lower left panel "inventory" to the "complete project" at the upper right panel is introduced and developed through design and photo dummy

38

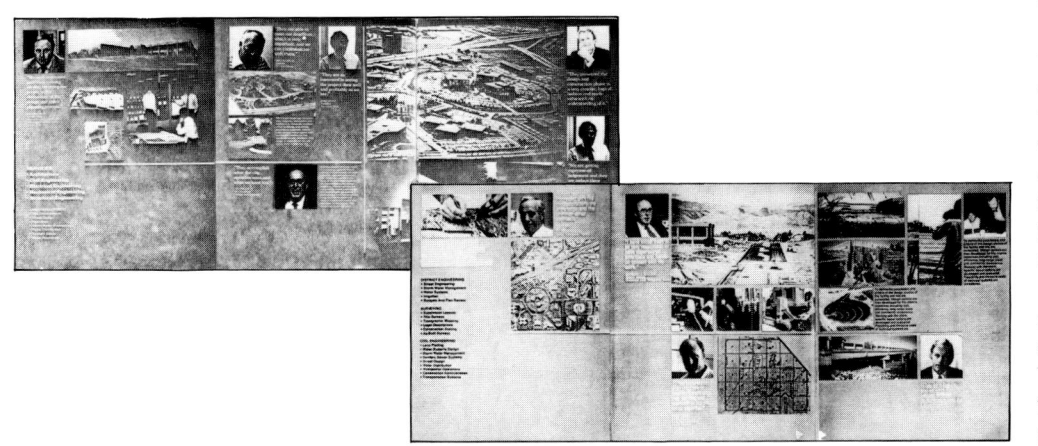

A corporate folder was developed for one of this firm's branch offices to hold a three-panel specialty brochure for a segmented market. One was designed for public clients and government agencies, and one for private clients and developers. The two brochures would never be sent together to the same client. Photo dummies were made using color xerox enlargements from 35mm slides and color polaroid prints, also enlarged from 35mm slides. This gives a close representation of what the completed pages will look like.

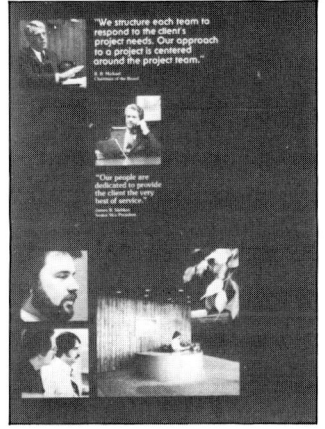

The commercial architecture folder is designed as an image piece using client testimonials and minimal project photographs. The matt black background provides sharp contrast to the colorful project pages. Studies were made to determine how the colored background would affect the readability of the type and how it would contrast with the photographs. The first, a light gray background with white type, did not have the impact the firm was looking for. The black field with white drop-out type made the type hard to read. The final choice was to make the drop-out type a soft color which provided a better contrast balance and looked more sophisticated.

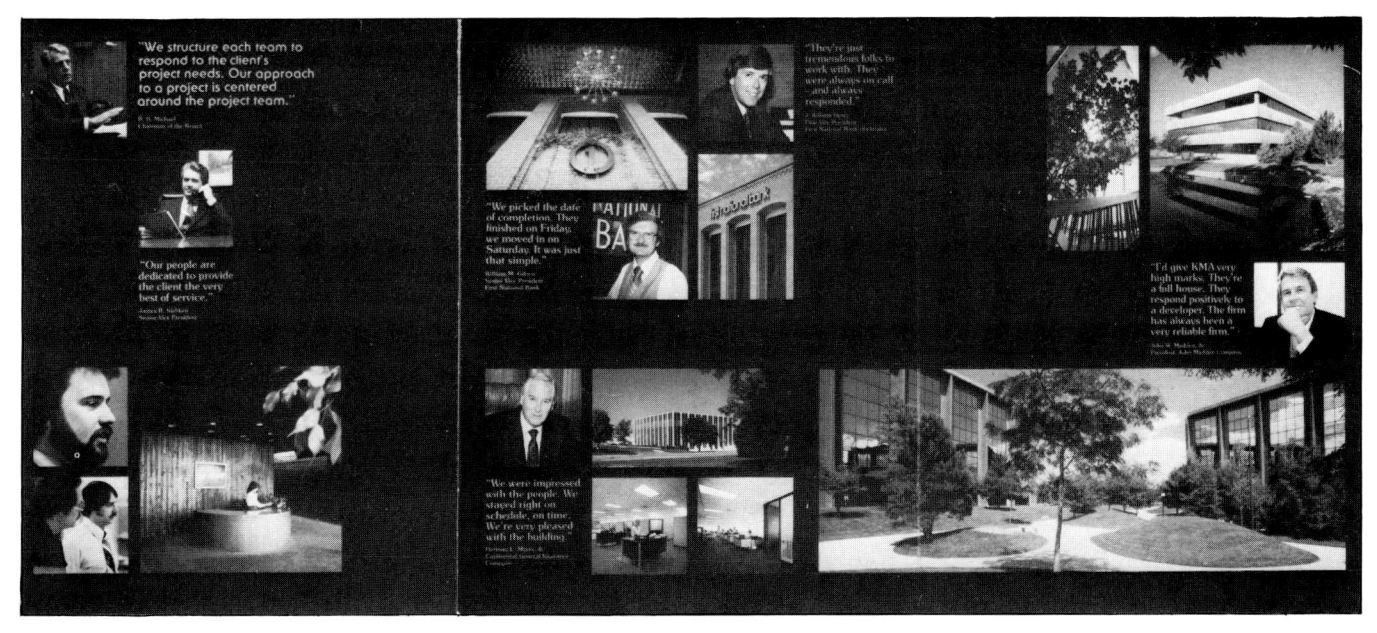

Designing The Format 39

Brochure Covers

Mid-State
This firm specializes in engineering, architectural planning, and surveying. The cover of its brochure features the logo, name and disciplines in white drop-out type. The photo used on the cover is a three-sided bleed close-up of mountains of paper work. An insert of the completed project is used, surrounded by a thin white border. The phrase "Concept to Completion" is used. This "concept to completion" idea is utilized throughout this piece with the use of full-page bleed photos and inserts to tell the story.

Clark Tribble Harris & Li
The preliminary sketch and finished cover of this architectural brochure features three of the firm's partners outside of their downtown office. The cover of this piece was designed to look like the cover of a newsstand magazine and the inside pages follow suit. The listing of feature articles in a magazine style type further fool the eye into thinking this company brochure is a magazine.

Michael Brandman
The aviation consulting firm uses a four-page foldout piece for its company brochure and direct mail advertising. The cover consists of a nose-on shot of a plane. The back cover is a tail-end photograph of the same plane. The front cover lists the company's name, disciplines, and company services.

Wright Pierce
The cover for this engineering brochure on water management features a full-page bleed photograph. Vertical column lines are dropped-out in white across this visual. The firm poses a provocative question on the cover which leads the reader into the brochure to find out more.

Sverdrup
The cover used on this engineering firm's specialty brochure shows highly polished, brightly-colored spheres placed on a black polished surface. The name of the firm and the subject of the specialty brochure are dropped-out in white type.

Greenhorne & O'Mara

This company's main brochure was designed to highlight its work in the areas of engineering, architecture, and planning, as well as in its other subspecialties. The cover shows the firm's home office (this visual spills over to the back cover too), with the firm's name appearing in white against the blue sky.

Zimmer Gunsul Frasca Partnership

This architectural, planning and design firm has presented a personable introduction on it's brochure cover through the use of people. Those shown on the front and back covers of this piece are architects and designers as well as principals with the company. The front cover presents a view of the inside of the firm's office.

Parkinson

This construction firm is situated in Alaska and the cover of its brochure gives a friendly but chilly first impression. The use of this humorous visual on the cover tempts the reader to open up this piece and read further.

Frizzell Hill Moorhouse Beaubois Architects

This architectural firm uses a deep deboss on the cover of its brochure. Each of the partner's names are stacked on top of one another and spaced accordingly to come out to the same line length. The style used in the deboss combined with the heavy-textured stock give the visual impression that the name of the firm had been carved.

Tishman

The cover of this four-fold piece used by this construction company consists of super high-contrast reversal photographs. Though all gray tones have been dropped out, the image of two construction workers working on steel beams still remains clear. The only color used on the cover is the firm's name and logo, printed in red.

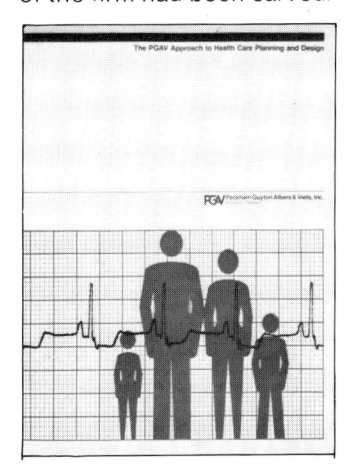

PGAV

This architectural and planning firm chose, for the cover of its health care brochure, a graphic design that represented a patient response chart, with graphic figures of a family superimposed over the chart. This graphic image relates particularly well to the specialty of the firm, which is health care planning and design.

Iowa Public Service

The cover of this public service company's annual report shows an embossed image of a house. The house's front window is highlighted in four colors, in an otherwise white-on-white image. The highlighted image shows a family in its home. It emphasizes the company's concern for its customers.

Sundt

This construction company chose for the design of its cover an aerial closeup of a construction site with a worker at the center of a web of steel rods. The only touch of color is the worker's red hardhat. Across the top of the piece, a black tab line and a white rule line separate the photo from the only type.

Designing The Format 41

How To Evaluate Your Brochure

If there is one item that every firm uses as a marketing tool, it's the firm brochure. And there is no item subject to more diverse opinion, or personal likes and dislikes and subjective aesthetic judgments, than the design office brochure. What makes one brochure stand out above another is sometimes elusive to define. Even more difficult to determine is what makes the most effective piece. With all of this subjective uncertainty in mind, we have developed an evaluation method whereby each element contributes to the total success. Obviously, there is no standard criteria established for good or bad design, but the brochure is more than an exercise in graphic design, writing, photography, and printing. Therefore, the function of the piece has to be the primary evaluation. In the marketing/sales cycle of contact/presentation/follow-up, the brochure is an effective means of establishing contact. Many use it as a means of presentation and of course, it can be a handy follow-up item. As a means of contact, the brochure is a substitute for personal contact. It must provide and produce an image, without any personal explanation. As a means of contact, it has most of its impact in the first 15 to 20 seconds. As the brochure lands on the client's desk, he picks it up and flips through it, sometimes from the back. If he is intrigued by some of the graphics, copy or photos, he may go back through it later, first reading only the headlines, then the bold captions, and finally the body copy. By this time he will have formed an opinion of your firm. The old cliche "you never get a second chance to make a good first impression" is very true here. The first impression is made primarily by the brochure's physical appearance.

PHYSICAL APPEARANCE

Cover
The first visual contact is the cover, and there appears to be two approaches one graphic and one photographic. On the average, the graphic design has less impact unless it is tastefully done. Full bleed photos have the most visual impact.

Color
Color usually enhances the look of each page, but black-and-white and color photos are hard to mix on the same page effectively. Poor quality color and black-and-white photos can also damage the image of a firm.

Paper/Ink
The overall look can be changed quite dramatically by the choice of paper and ink, but usually for the worse if you stray too far from black photos on white paper. There are scores of shades of white paper, for example, and a wide choice of paper weights. Slightly textured paper tends to reduce the clarity of half tone photos.

Design
Photo quality and composition are all subjective integral elements which contribute to the look; but probably are more a factor to the designer than the client.

Binding
The horizontal brochure is difficult to hold and leaf through due to its 8½" x 22" inch format. Also the wide dimension is outside of our normal field of vision so that each page becomes its own focus, rather than the 11" x 17" inch double spread, the most comfortable to look at.
The stapled (or stitched, as it is called in the trade) binding seems to be superior to a plastic spiral wherein the comb interferes with the clean look. The plastic spiral binding is the most acceptable method of customizing material to a specific client. Loose pages are usually just glanced at and tend to get out of sequence easily. Also, they can fall out of the folder.

Illustrations/Charts
These must be high quality in terms of draftsmanship and must also be easily comprehended. A poor graphic or sketch illustration that looks amateurish will drag the visual appeal of the whole page down in its level of quality. The same is true of architectural renderings.

Text/Graphics
If the page looks crowded with text, no one will stop long enough to read it. If color ink is too faint it will be skimmed over. People read the easiest things first. Clean simple type with ample line spacing is the safest. Tests have shown that ragged columns are easier to read (on the average) than justified columns. Sometimes justified type looks more appealing when running alongside a photograph.

QUALIFICATIONS

History Of Firm
This means simply a clear description of the firm's background, not a detailed blow-by-blow of the evolution from the founding father. The amount of detail should be in proportion to the strength that the information lends credence to the objectives of the brochures.

Services
Often overlooked and often misleading, particularly in the case of consultants and constructors. The safest bet is to relate the services to the process not the product.

Projects
Many brochures try to show too many to illustrate diversity, which only tends to clutter the pages.

Text
Brochure text must be clear, concise, and oriented towards benefits to the client, otherwise it will work against you. Avoid "we" oriented material.

Brochure Evaluation Form

		COMMENTS	PTS
PHYSICAL APPEARANCE — 3 pts. max.	VISUAL IMPACT: Eye Grabbing Pleasing Mediocre Plain		
	COVER: Eye Grabbing Pleasing Mediocre Plain		
	BINDING: Easy to Open Attractive Distracting Poor Choice		
	DESIGN: Outstanding Very Good Mediocre Poor		
	PHOTO QUALITY: Excellent Acceptable Mediocre Poor		
	PHOTO COMPOSITION: Excellent Good Inconsistent Poor Mix		
	PAPER / INK: Good Quality Good Balance Fair Poor Choice		
	COLOR: High Quality Effective Good Use Waste of Money		
	TEXT: Good Balance Page Crowded Too Weak Not Enough		
	GRAPHICS: Outstanding Well Designed Acceptable Intrudes on Photos		
	ILLUSTRATIONS: Appealing Good Mediocre Poorly Done		
	CHARTS / DIAGRAMS: Understandable Appropriate Confusing		
MARKET IMPACT — QUALIFICATIONS 5 pts. ea.	HISTORY OF FIRM: Well Portrayed Overdone Not Enough Information		
	EXPERIENCE: Clearly Demonstrated List Clients Only No List		
	SERVICES: Applications Well Described List Only No List		
	PROJECTS: Well Represented Too Many Poor Mix Too Few		
	KEY PERSONNEL: Good Image Too Many Unimaginative Not Shown		
	TEXT: Informative Well Written 'We' Oriented Too Wordy		
	STAFF: Appropriate No. Good Display Too Many Too Few		
	TAILORED TO CLIENT: Excellent Too General No Specific Focus		
	SEGMENTATION OF MARKETS: Good Acceptable Fair None		
	UNIQUE SELLING POINTS: Many Some Few None		
	DIRECTED TO NEW BUSINESS: Yes Somewhat Record Only No		

OVERALL EVALUATION

OVERALL IMPRESSION: 10 pts. max.
Exciting Pleasant Good Fair Poor

A. PHYSICAL APPEARANCE 35% .... ____

B. INFORMATION ON QUALIFICATIONS 35% .... ____

C. MARKETING IMPACT 20% .... ____

D. OVERALL IMPRESSION 10% .... ____

TOTAL POINTS .... ____

RECOMMENDATIONS:

DESCRIPTION	RATING:
Excellent Piece	100%
Superior Quality	90%
Very Good	80%
Good	70%
Needs Updating	60%
Needs Revision	50%
Start Over	40%
Poor	30%

Designing The Format

Brochure Formats

Key Personnel/Staff
Many brochures do not show principals, much less key personnel. The most common reason being the turnover among people and the potential "dating" of its shelf life. Less valid reasons extend to "our people are too young," or "people with beards," for example, "might prejudice the client."

Tailored To Client
Keen marketers have given up trying to impress the client with diversity and are focusing on specific needs of individual clients.

Unique Selling Points
If a brochure is just a record of past projects there is not much to distinguish it from competitors. One unique selling point used to be to show computer capability. It's not unique now to show computer capability.

Directed To New Business
Some may feel that by showing a picture of a building with no identification and no text it illustrates the fact that you've done that kind of work. A brochure which features a combination of process along with a record of projects is more successful.

Overall Impression
After you've taken a super-critical look at your brochure, how you feel about it can give you some extra points.

BROCHURE EVALUATIONS SUMMARY
Each year promotional material of leading design firms is judged and evaluated by various organizations. The material consists of simple foldouts to sophisticated brochure systems. The exact breakdown was as follows:

90-100%	Superior	6%
80-89%	Very Good	10%
70-79%	Good	12%
60-69%	Needs updating	33%
50-59%	Needs revision	21%
40-49%	Start over	14%
30-39%	Poor	4%

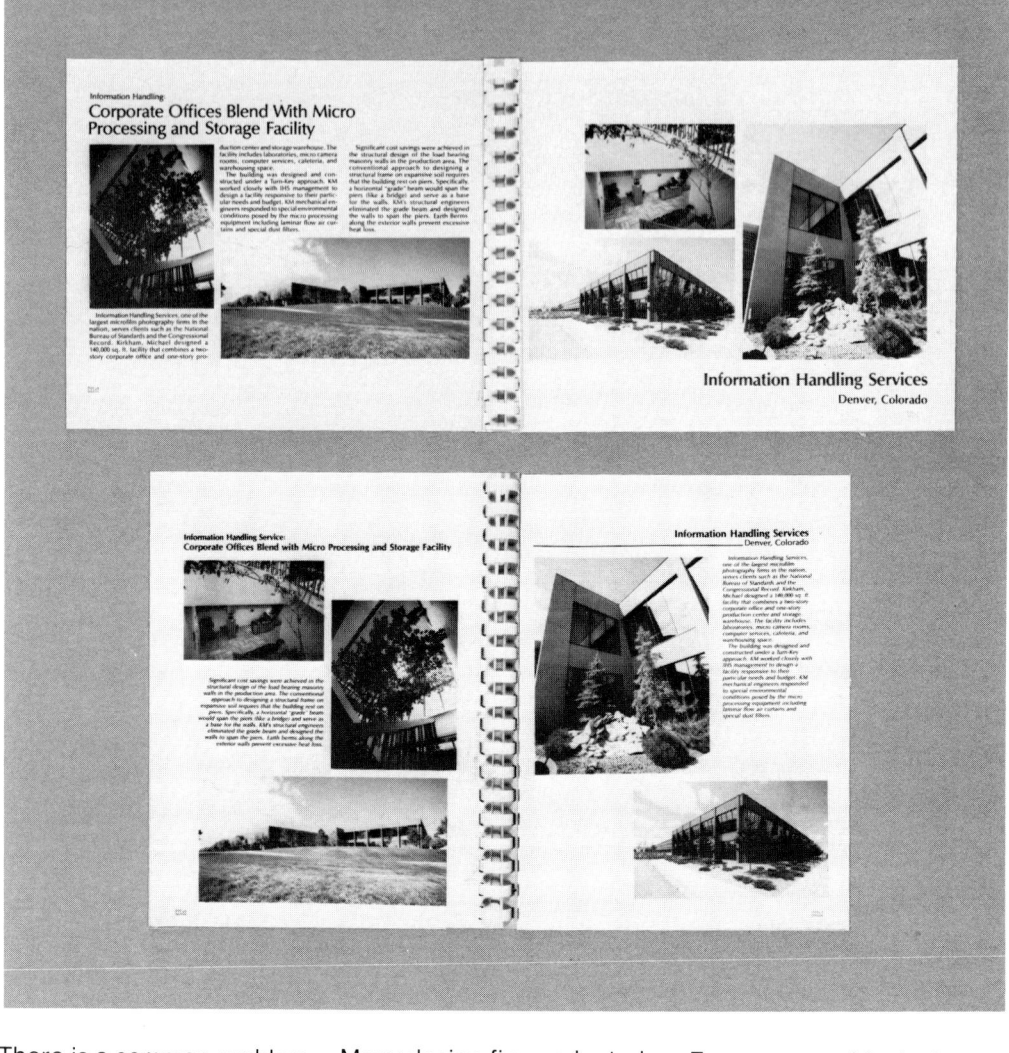

There is a common problem among many A/E firms regarding the proper format for brochure, proposals and technical reports. This problem is the choice between a vertical or horizontal format. The horizontal format is called various other names such as: "oblong, portfolio or landscape." The vertical is commonly referred to as "standard," and for a very good reason. It is the most commonly preferred style of newspaper, magazine and book design. Why then should the architect or engineer stray from this standard format when designing their promotional material?

Many design firms adopted the horizontal format just to be different from the rest. They soon discovered that many other firms were also using it, so they abandoned it and went back to the standard vertical format. The illustration above shows a typical page layout using the identical size photographs for each, one horizontal (top) and one vertical (bottom). While both layouts are pleasing, the vertical pages focus attention much better and get the message across much quicker. The double spread provides a comfortable horizontal image that is easily scanned.

For years graphic design consultants have been recommending the use of horizontal format slides and vertical format brochures. If this sounds inconsistent, consider an open 8½" x 11" inch book or brochure, it measures 11" x 17" inches, almost exactly the same aspect ratio as a horizontal slide (1:1½). The reason for the horizontal format slide, aside from the fact that vertical slides almost never seem to fit on the screen, is due primarily to the cone of vision of the human eye. Each eye has an elliptical cone of vision of 40° vertical and 100° horizontal. Since humans have what is called binocular vision, the cone of vision of each eye overlaps producing an ellipse of 20° vertically and 45° horizontally. This cone of vision is very close to the aspect ratio of a horizontal slide or an open double-page spread of a book. Good book designers know that each spread has to be treated as one image and the same is true of brochures.

Graphic design and communications consultants have often commented on the use of the horizontal or the vertical format. One graphic design consultant felt that the horizontal format is appealing because it is different, although people seem to be more comfortable with the vertical format. A corporate design consultant stated that it is a vertical world we live in - graphically. Because of the broad range of applications that brochures and marketing materials have to meet, the vertical format is far more flexible and applicable to those various requirements. A brochure designer specializing in architectural and engineering brochures feels that when you open a horizontal brochure the 22" inch length becomes unwieldy. A communications consultant responsible for the design for many architectural and engineering brochures recommends sticking to the vertical format. From a very practical point of view, considering paper stock sizes, and press runs, you end up with alot of waste when you try to run a horizontal format. Coated stock can only be folded one way. These and other comments give evidence to the inappropriateness of the horizontal format for brochures. Consider a typical magazine stand, the horizontal format is completely unknown.

Promotional Material Vocabulary

Corporate Identity Program
The graphic representation of a firm to a variety of publics which, at the same time, fulfills regulatory, operational and marketing needs. It is applied to the entire package of print materials used by the firm or company.

Corporate Advertising Program
Individual advertisements placed in business, trade or general-intrest publications. A campaign usually consists of three ads with a creative related concept.

Communications Program
A comprehensive communications program using a variety of media in a single campaign or in the firm's total marketing program, showing a consistency in the development of advertising, brochures, newsletters, media placement, and direct mail campaigns.

Annual Report
Reports made annually for distribution externally for publicly held or privately owned companies. Usually focuses on accomplishments of the company, or firm, during the past year, progress, future expectations, and financial report. The financial information is usually in a totally separate section from the general information.

Direct Mail Campaign
One or more packages of marketing materials developed to reach a specific target audience and to elicit a response. It can include circulars, fliers, form letters, pamphlets, response mechanisms, announcements, newsletters and market specific brochures.

Special Events Piece
A printed piece or unusual item representing a one-time effort for a special event. This can include anniversaries, a new name, new office, a holiday, staff promotions and relocations. It can also include work produced for a client such as ground breaking and dedications.

Company Brochure
A general corporate publication giving introductory information on the firm's total capabilities.

Special Market Brochure
A publication representing a one-time effort to market a particular service, discipline, office, or joint venture arrangement to a targeted market segment.

Service Brochures
These are intended to introduce the firm or a specific part of the firm to a potential client. They outline specific services to specific markets.

Category Brochures
These are usually devoted to a specific building type with text related to the depth of experience in the project type.

Magazine
An external publication with liberal use of photos and art, and usually employing a more interpretive writing style. The format is feature oriented.

Newsletters
A high-frequency external (or internal) publication with few pages, employing a more concise writing style. The format is news or news feature oriented.

Pocket Folder
A folded sheet of heavy stock designed with a pocket on one or both sides of the inside panels. The pockets are glued to provide a method of holding supplemental loose material. Brochures can be designed with a pocket folder on the last inside page. They are usually die-cut for the placement of a business card.

Project Page
A single sheet printed on one or both sides that can be arranged and bound up in any sequence or used loose in a pocket folder. Usually focusing on the visual aspect of a project.

Project Fact Sheets
One sided sheets focusing on project data such as project name, client, square footage, and other technical data to explain the functional aspects of a project. They sometimes contain a floor plan or project site plan but rareley include photographs as the main focus.

Case Studies
These concentrate on a single project that has specific significance and application to other projects of its type.

Resume Pages
Usually one sided single sheets with an individual resume, with or without a photograph.

Client Lists
The mainstay of any design firm is its list of clients. The listings can be categorized in several different ways, either by project type or by client type. Clients within each listing can be arranged by order of importance, by order of date of completion or alphabetically.

Designing The Format 45

FORM & RESPONSE 2

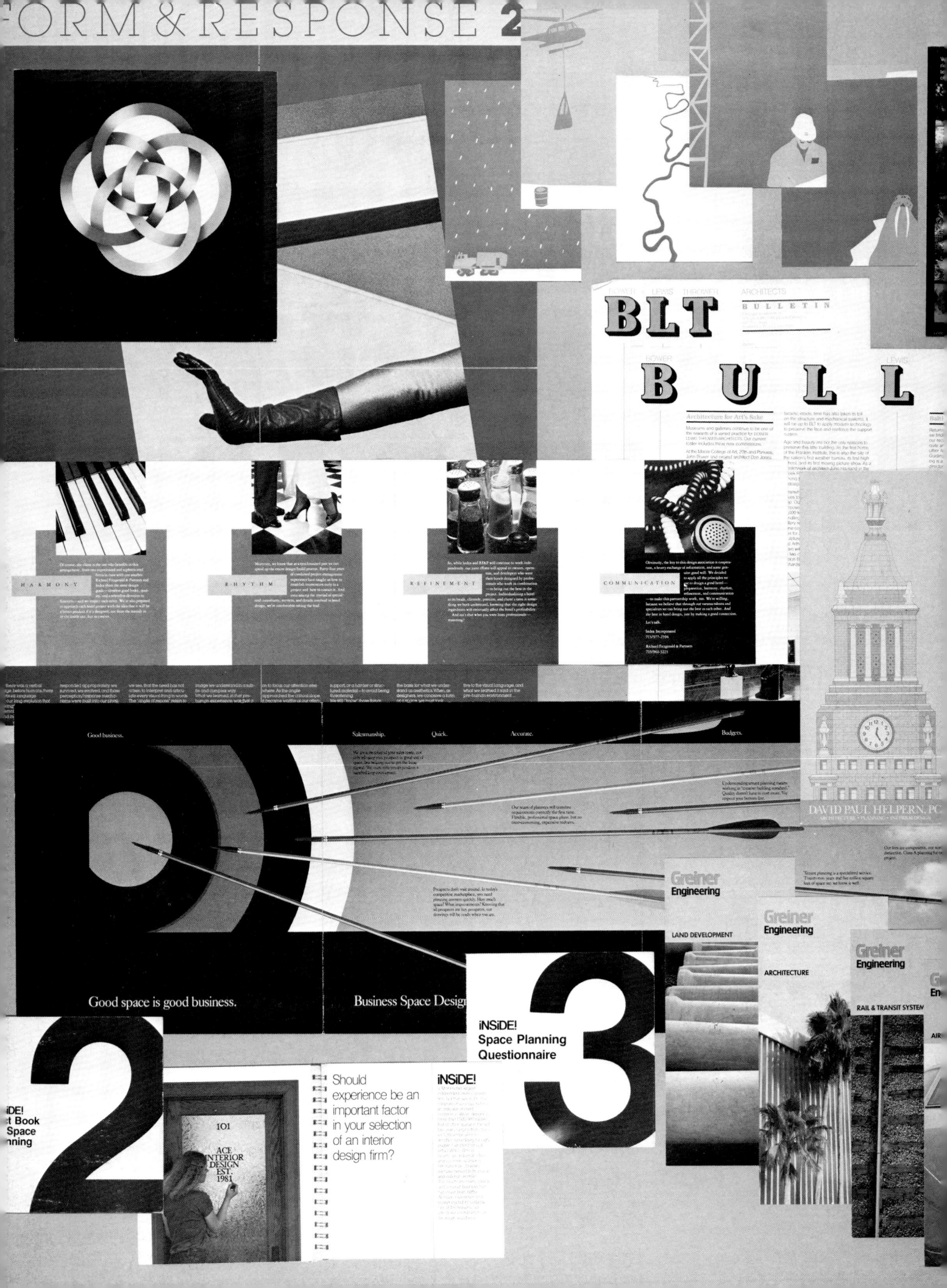

DIRECT MAIL: 3

Index

The goal of this promotional campaign was to produce a timeless quality piece illustrating an analogy between the goose which laid the golden egg and the firm, both standing for handsome, functional, natural design. The piece was meant as a "door opener" to be followed later by the company prochure and accompanying pieces. The format for this main piece is an accordion foldout, showing geese, text, and finally the golden egg.

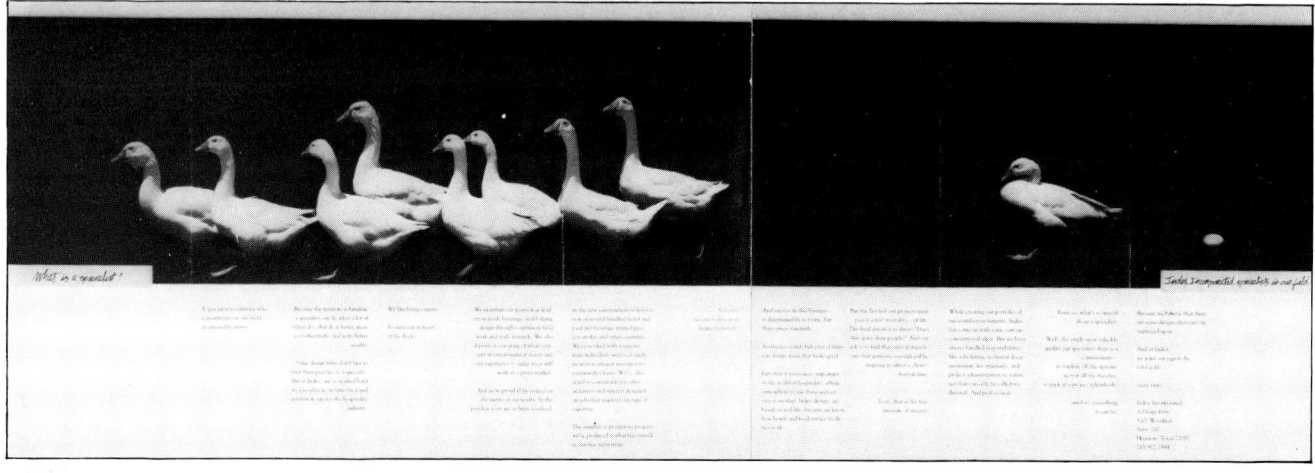

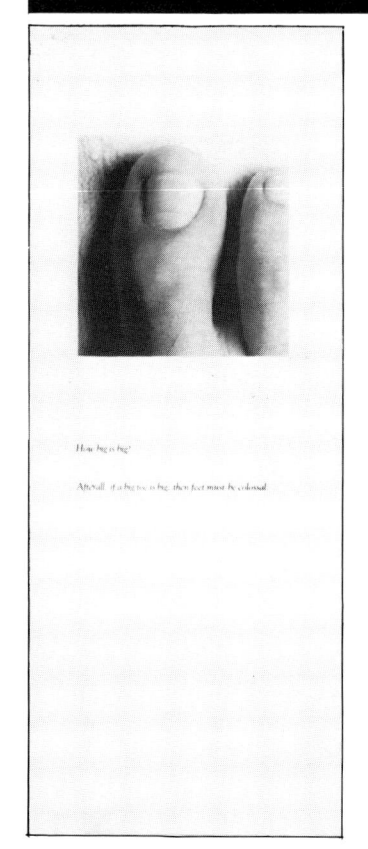

Pierce Goodwin Alexander

The two direct-mail pieces created by this architectural firm measure 4½" x 11" inches when fully closed. These pieces open up to seven panels, folded accordion-style. The first piece discusses how large the firm is. The photographs on each panel relate to "big" things such as a "big" toe. The second piece discusses the qualifications of the firm in a non-specific manner. The photographs relate to the text in a similar manner. One example: the text discusses how hard it is to find a design firm with the right qualifications and the photographs on these two panels show, first, a needle and, second, a haystack. Other familiar sayings are illustrated, such as: a silk purse, a sow's ear.

Pan Am CPS

This interesting foldout (in black and white) shows a helicopter over the New York City skyline. There are three black tabs with minimal text on the right-hand side of the piece. When the customer holds the right and left side of the foldout (and pulls), it opens to reveal a full spread that describes, visually and with text, what the company does. When the piece opens, the helicopter is divided into three parts (end, middle, and front). These parts are seen on the different panels of the piece, and remind readers of the speed with which their packages will be delivered. The two-tone format (beige and white) is accentuated by the black tabs and white type. The company logo also appears.

Charles Caplinger

This piece folds up to fit into a number 10 envelope. The black-and-white foldout displays the company's name and its logo. Individual panels are divided by black rule lines. Each folded panel section shows a visual of either a site plan rendering or a photo, depending on the stage of completion of the project. Vital information about the project is limited to one panel. The foldout also includes a list of projects, clients, and the firm's areas of expertise.

Direct Mail 49

Ellerbe

This multi-disciplined firm has created a direct mail package that makes it extremely easy for the potential client to respond and receive further information. A series of three 5½ inch x 5½ inch booklets is sent to potential clients packaged in an easy-to-open box. The boldly numbered booklets are attractive and inviting to read. Booklet one is spiral bound and uses full-color illustrations and photographs along with text. Booklet two uses various type styles in many colors for its graphics. Booklet three is an accordion folded piece and its text consists of a yes or no questionnaire. Pieces one and two provide pockets at the rear of the booklet containing a postage paid reply card. The card in booklet number three is a perforated tear out.

Specialty Brochure

A horizontal format brochure is sent to potential clients as a response to the booklet reply cards. The brochure offers nine design and space planning solutions as discussed in the mailer pieces.

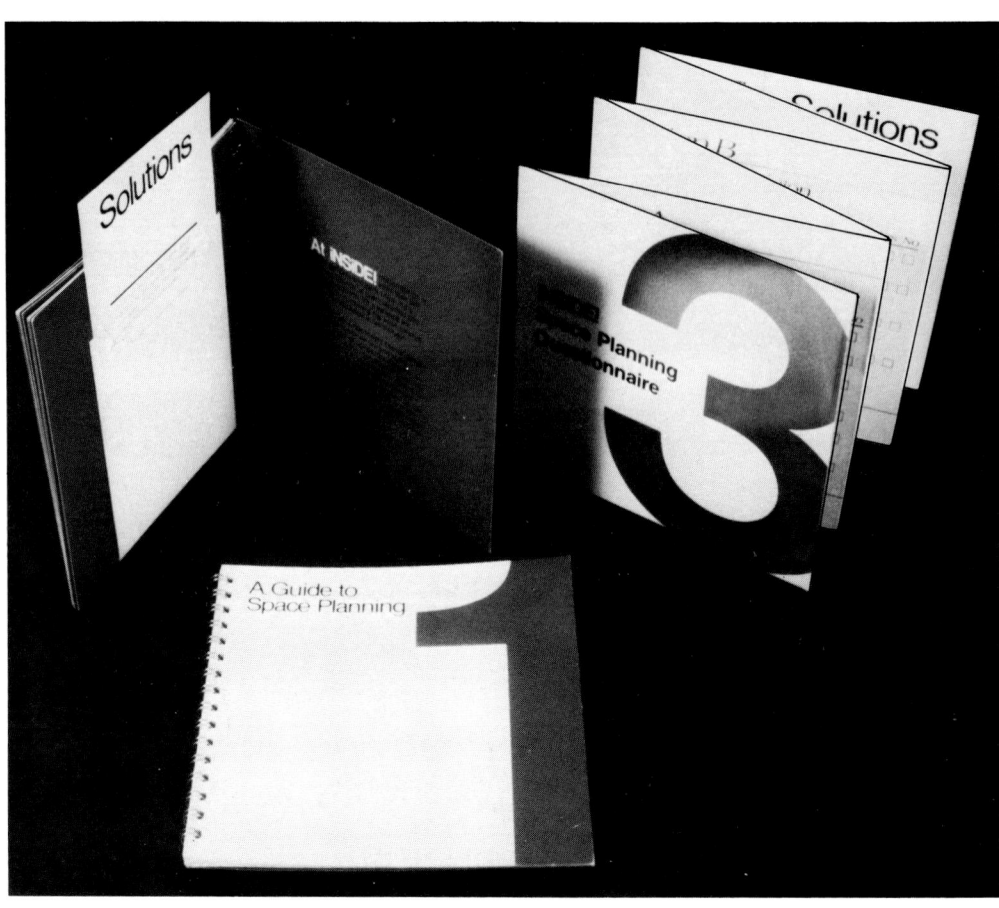

Business Space Design

Commercial office space developers were targeted for this direct mail campaign of seven thematic postcards, which were mailed out at two-week intervals. The themes were selected to respond to information gathered from an image survey this multi-disciplined firm conducted. Poster-size enlargements of each card were displayed at events and client presentations. Additionally, a fold-out four-color brochure was created. This piece was the culmination of the mailing series. Across the top of the poster/brochure are thematic headings that correspond to the individual postcards.

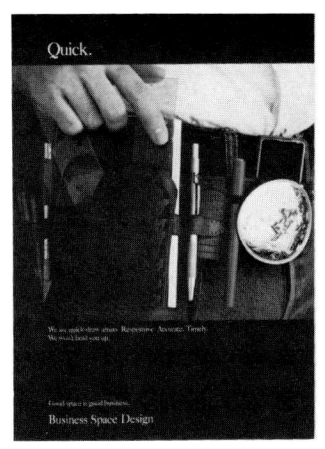

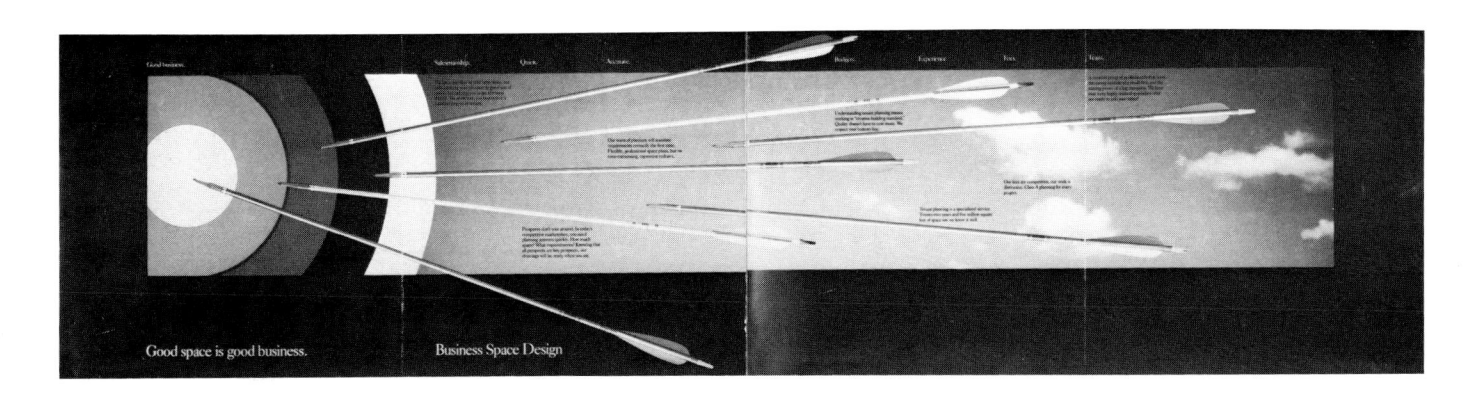

Direct Mail 51

CMC

This construction firm has two brochures, one of which features a silver field with red rule lines and red-and-black type. The front cover shows a black-and-white photograph of a blueprint. Inside, the company's services and clients are listed in separate sections. The other brochure has a picture of a concrete structure (in a mezzotint), and is printed in gunmetal tones. The red rule lines and black-and-red type are maintained. Inside, the text discusses the firm's expertise in management and its overall expertise in construction. This piece includes black-and-white photographs and charts, as well as a list of services and a partial client list.

ADD

This architectural firm developed its brochure to inform clients about the importance of integrating art and architecture. The four-color glossy front cover shows a building with a contemporary piece of sculpture.

Raymond Hansen

This piece clearly conveys this landscape architectural firm's expertise. Front and back covers feature a floral design in rich colors. Inside, four-color photos are placed against a glossy black field. The text is limited to photo captions. The firm's multiple services are listed across the bottom of the page.

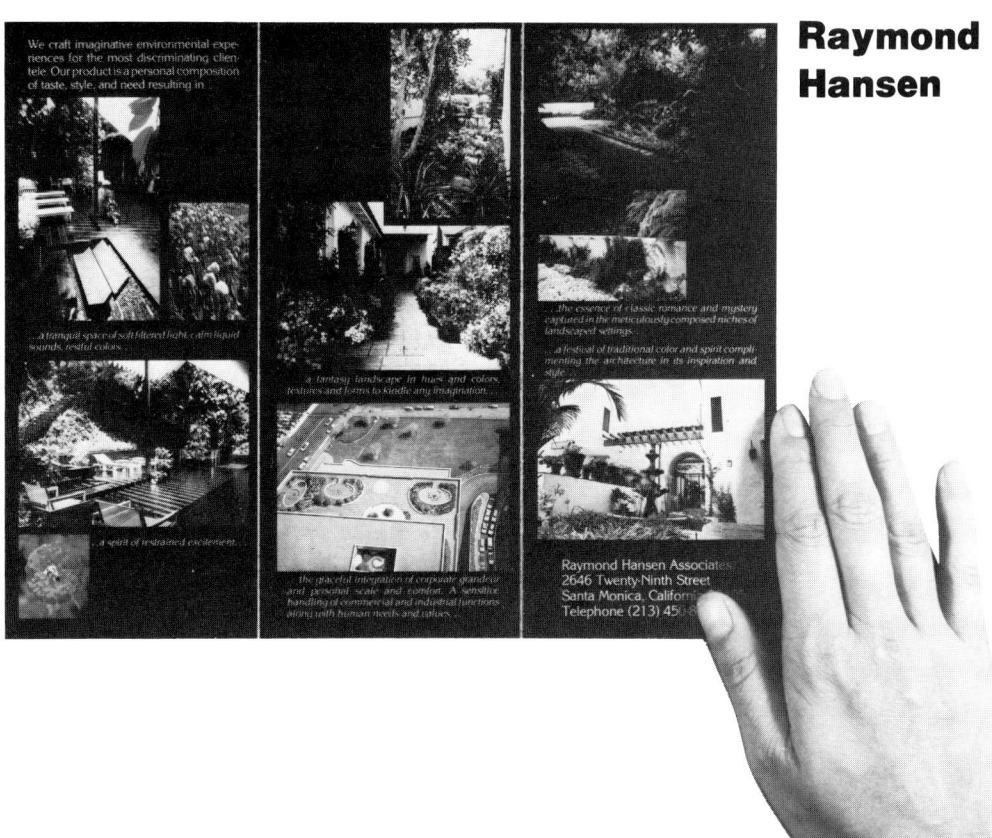

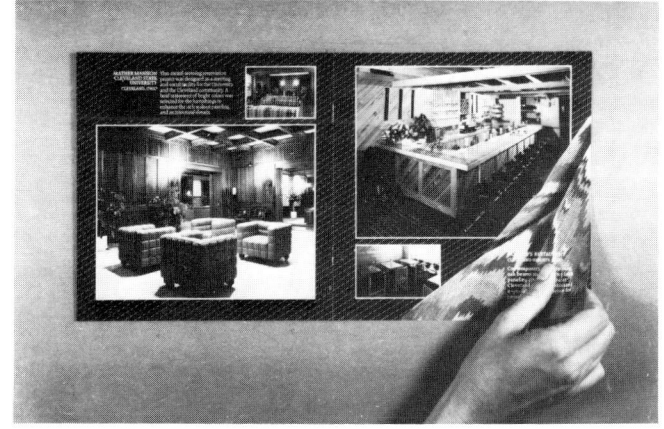

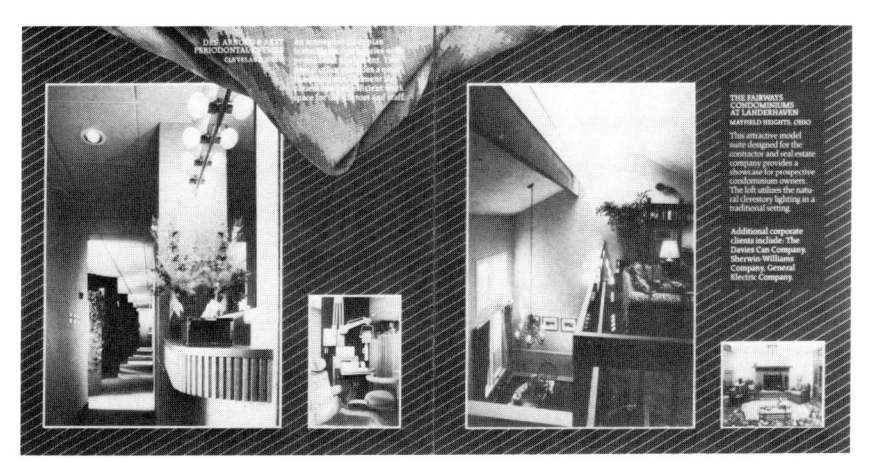

Helen M. Moran

This interior design firm has created its marketing materials to act as a partial presentation of style and quality. The cover shows a photograph of a commercial interior in four colors. A piece of cloth is shown across the front of each photographic spread, to accentuate the interior design aspect of the brochure.

Direct Mail 53

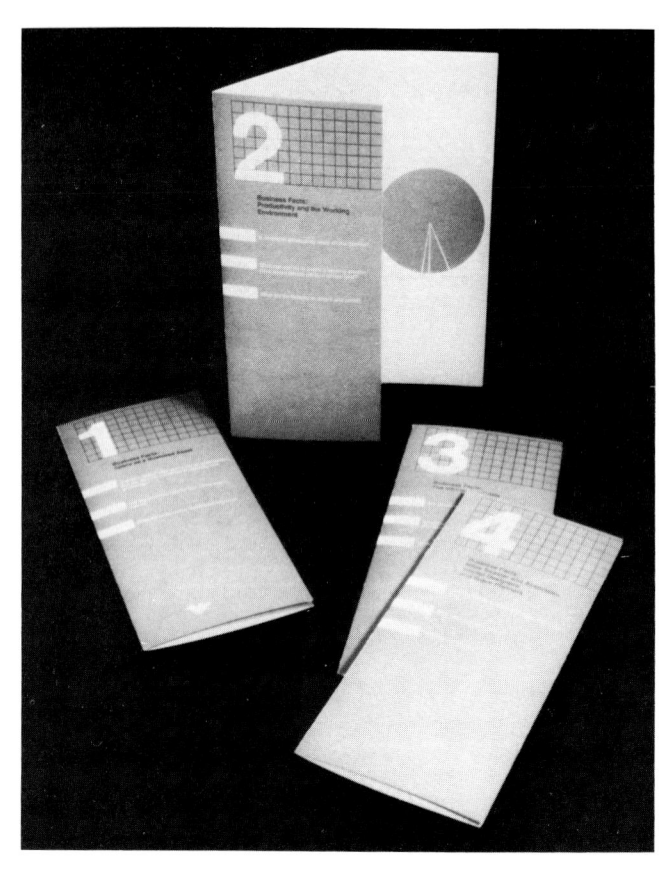

Miles Treaster

This planning firm designed a series of gray, red and white fold-outs to highlight the range of projects with which the firm is involved. On the front of each piece there are white tabs with red numbers. These tabs are cued into the interior information.

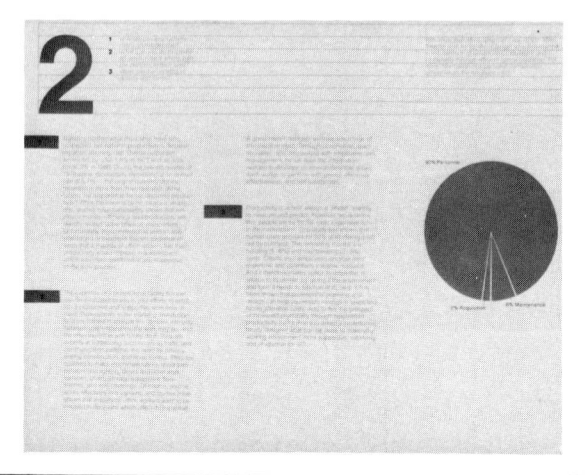

Brae Construction

This self-mailing fold-out piece (in blue and white) opens first to reveal four-color photos of the firm's personnel. The text describes the company's approach to construction and how it resolves building problems. At the bottom of the piece, a panel opens again, revealing the firm's logo, business card, and business reply card. The back of the foldout includes four-color photographs, the name of the firm, and its logo.

Graham/Meus

This architecture firm designed a self-mailer to announce the opening of its new location. The long, narrow foldout mailer includes a removable standard Rolodex card with the new address and effective date on the reverse. The logo repeats so that once the card is removed, the name of the firm remains on the piece (the piece is red and white with black type). The variety of black-and-white illustrated projects and project photographs furnishes a well-rounded survey of the firm's expertise and diversity.

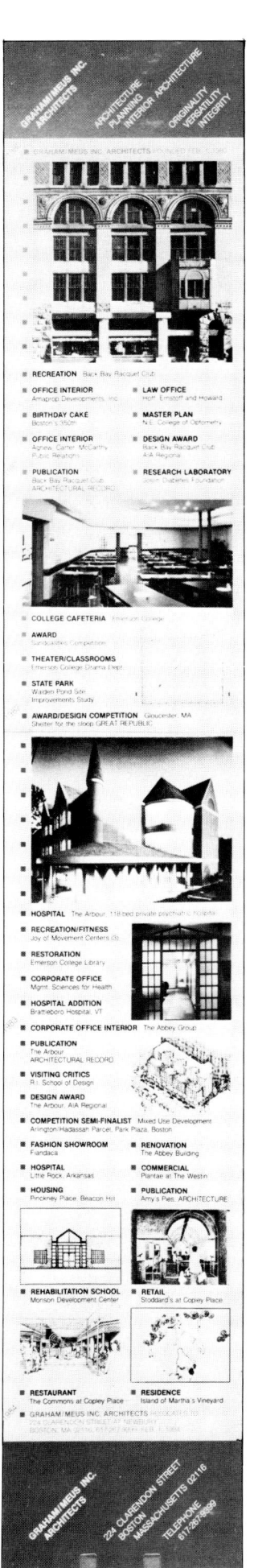

Downing Leach

This architectural and planning firm has designed a vertical foldout piece as a marketing reminder that the firm has the expertise a client is looking for. The colorful and contemporary foldout features a graphic on the cover. The border and the type are both red, with the graphic shown in blue, gray, black and white. Sections of the graphic are repeated on the inside, and built back up from panel to panel. The last section on the piece is similar to the cover except the colors have changed. When the piece is turned over it is a solid red with the name of the firm and its address in black type. This piece can be used as a self mailer.

Direct Mail 55

Ewing Cole Cherry Parsky

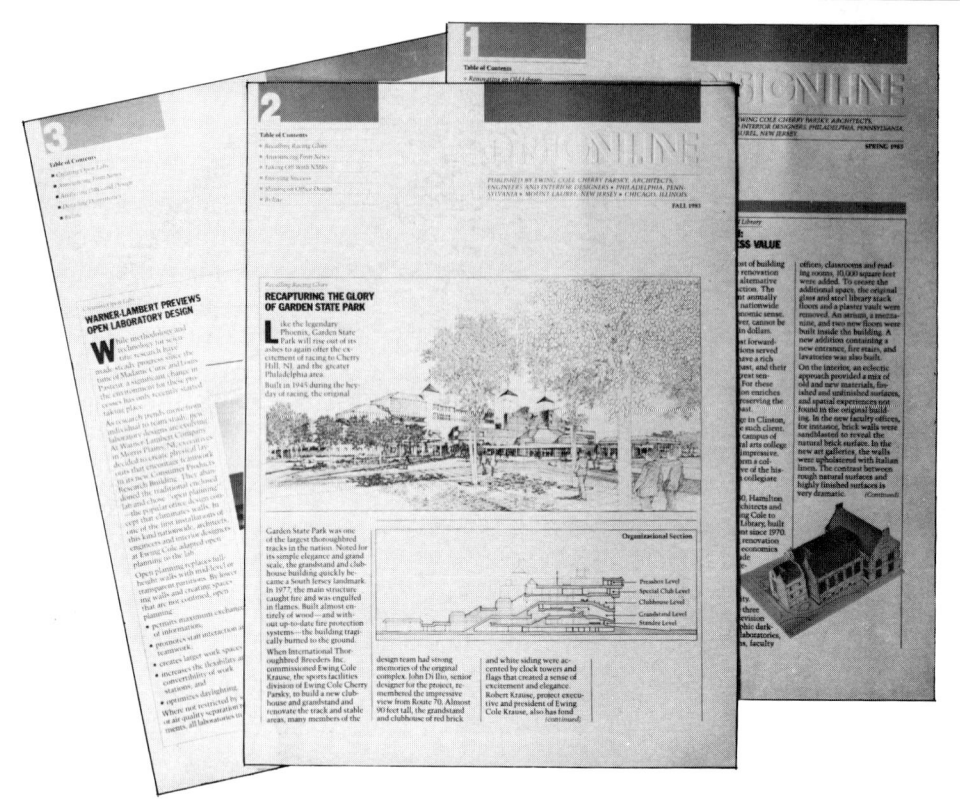

This multi-disciplined firm has created an oversized (9½ X 11) three-fold format for its newsletter. A four-page newsletter is also used as an alternate format and it, too, is oversized. The cover design is kept consistent, with only the number of the newsletter and the articles changing from issue to issue. The name of the newsletter, "Designline," is conveyed on each cover as a white-on-white blind emboss. This newsletter's layout creates an airy feeling through the use of two colors and the design decision not to "fill up" every inch of space available. The four-column format and the use of subtly shaded renderings and photographs make this newsletter easy to read and easy to review.

Yearwood & Johnson

This architectural firm uses their newsletter as a cost effective means of establishing or solidifying client relationships. This newsletter is scheduled for publication three or four times annually and carries news about the firm's performance in various market segments. A different segment is highlighted in each issue. Project matrices and building diagrams are part of the interesting graphics

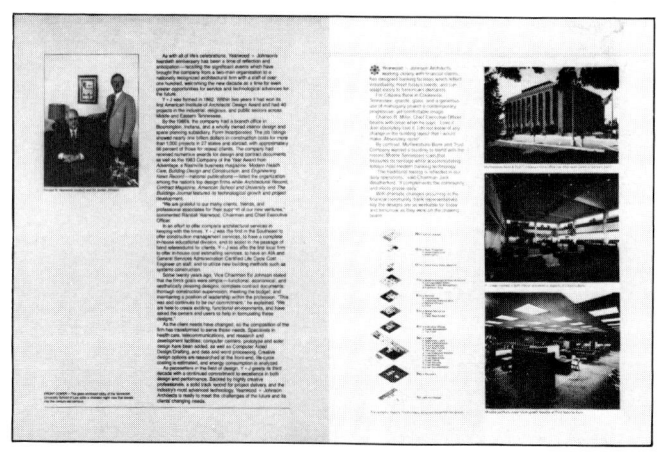

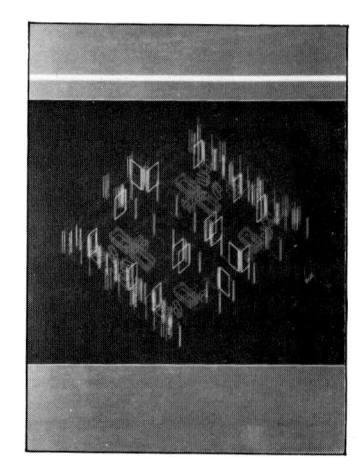

Bower Lewis Thrower

This architectural firm publishes a newsletter, as their main promotional vehicle. The piece is designed with an accordion fold, revealing the first initials of the partners' surnames. When opened, these same first initials (BLT) become a part of the word BULLETIN. The piece is laid out in six panels with colored rule lines that change from issue to issue.

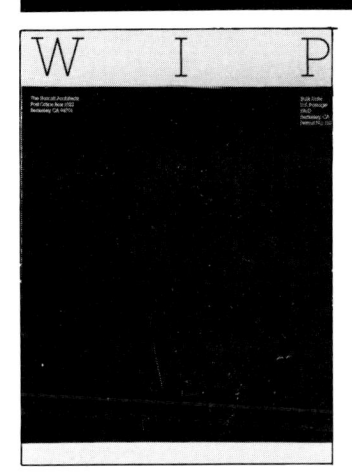

Ratcliff

This two-color self-mailer newsletter is 8½" X 11" when mailed and unfolds to a 17" X 22" size. The format is kept the same from month to month. The initials "WIP" appear on every issue and stand for "works in progress." One side of the newsletter discusses recent projects and the other side is a full-page photograph. Though this poster-size photo is changed for every issue, the subject matter is usually limited to a shot of a recent project with a contrasting image included. Brief text on the poster side discusses company philosophy. The words "Form and Response" appear in each issue, printed in the colored strip at the top of the page. The color of this strip is changed from month to month.

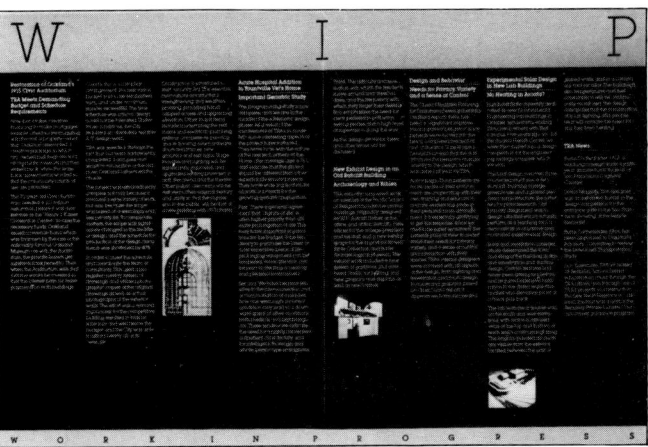

Direct Mail 57

Hammel Green Abrahamson

This newsletter, designed as a direct mail piece, is sent to upper-echelon management two or three times a year. The newsletter was designed to be browsed through by the busy executive. For this purpose, there are large color visuals and little text.

LEA Group

A color photograph centered within a grid background with colored type is the consistent cover design for this quarterly letter. However, the color choice and photograph changes from issue to issue, featuring the firm's most recent accomplishment. The grid pattern is repeated on the inside of the piece.

EDAW

This landscape architecture firm's newsletter uses a two-color, eight-page format. Black-and-white visuals are used, with the text kept to one topic per issue. Bold black type and gray rule lines are prevalent. The fold-out accordian design of the piece was created to be bound in with the firm's looseleaf format brochure. The accordion folds cover up all personal information when bound, allowing it to also be used in proposals. The piece uses a horizontal format designed to pull the reader into the piece. Visuals include aerial views, interiors, renderings, models, and completed projects. A middle band of gray is repeated throughout the piece; within this band, relevant quotes from various sources are printed. The back cover carries an excerpt from a current article by or about the firm. The newsletter is directed toward former, present, and potential clients.

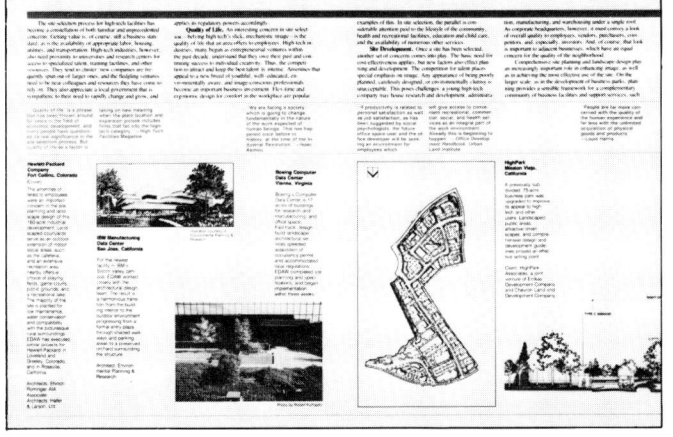

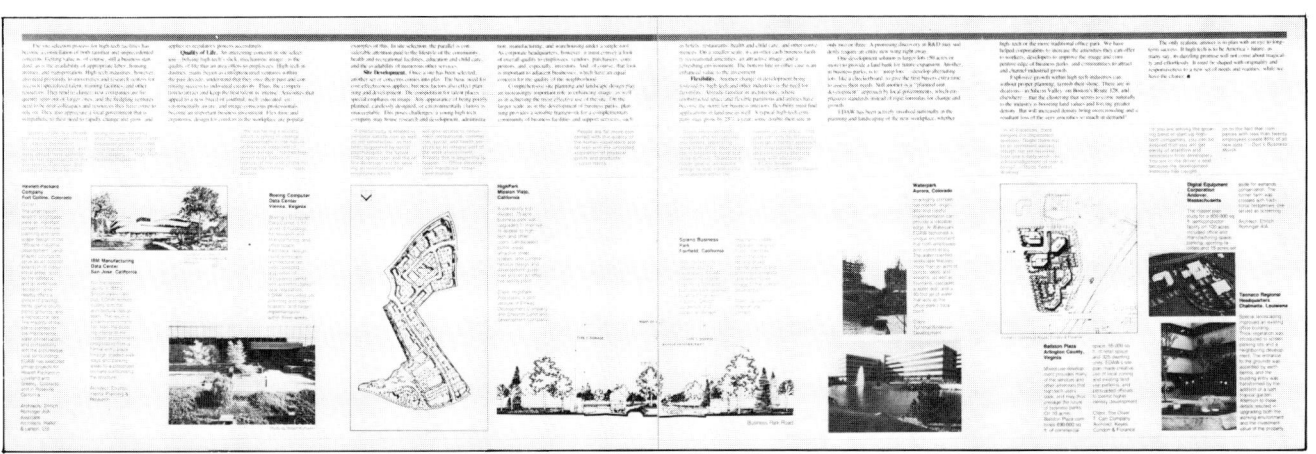

Direct Mail: Newsletters 59

Jones

This firm targets its magazine to construction people, potential clients, joint venture partners, financial institutions, and the firm's employees. Articles in the magazine are written to teach and inform readers about the firm's involvement in the construction industry. The journal is published three times a year, and the cover design is consistent from issue to issue. Four-color photographs or illustrations are used with a white rule line bordering the top of the visual, and white type. All issues include a letter from the president as an introduction to the reader. The back cover includes a twelve-block grid of photographs taken from the inside of the magazine.

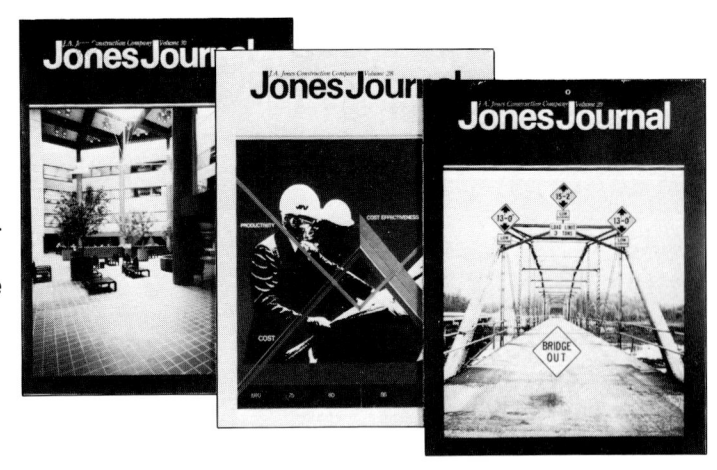

Enterprise

The covers of this firm's magazine are in four colors against a dark background, and are enclosed in a white rule-lined box. This design is maintained throughout all issues. The back covers show people and projects, and include quotes and text. Inside, subscription inserts are bound to the binding. The paper is strong glossy stock.

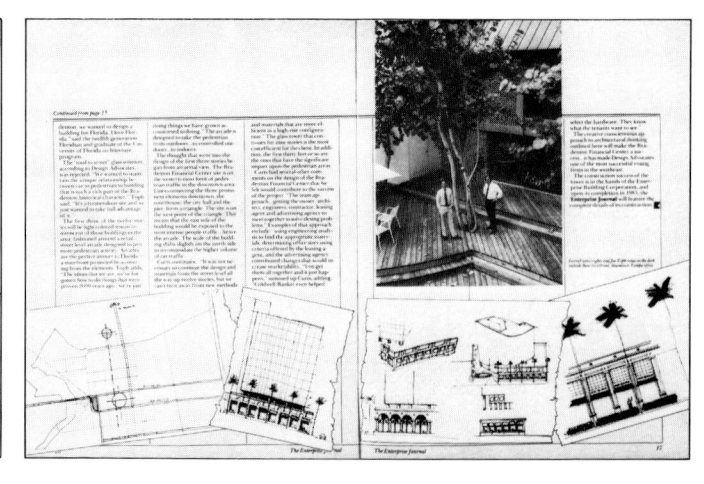

60

The Maguire Group

This four-color magazine stresses the architectural aspect of the firm in order to emphasize the company's work in that area, since it is well known for its work in engineering and planning. The covers are all in four collors, with strong visuals and bleed photographs. Quotes and photographs of company members contribute to the personal tone of the magazine. The text includes detailed descriptions of particular projects, and discusses the firm's approach to architecture, engineering, and planning. Articles usually focus on specific projects, international firm-related news, and upcoming opportunities for the firm. Although the design changes from issue to issue, the logo and type on the cover remain the same, thus creating a consistent appearance.

Direct Mail: Magazines 61

Post, Buckley, Schuh & Jernigan

A full-color advertisement for this multi-disciplined firm use a simple and eye-catching approach on its visuals. Employees are shown using the technology of a CADD System. The text is kept simple and discusses the firm's capabilities.

Myklebust Brockman

This multi-disciplined firm has produced several ads that use the same theme. Commonly-seen items are used with related text, such as a photo of burnt toast and the opening line, "Don't scrape burnt toast." referring to the fact that correcting errors doesn't match doing it right the first time. The slanted type lends itself nicely to the simple design and the use of the logo and address make this ad a complete information package.

Lehrer/McGovern

This construction/management consulting company's advertisements are designed to appeal to builders, architects, developers, and corporate facilities people. The two-page color-spread ads can also be used as marketing pieces. All photographs are numbered and keyed into the text.

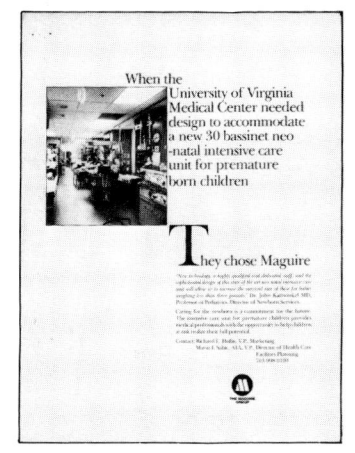

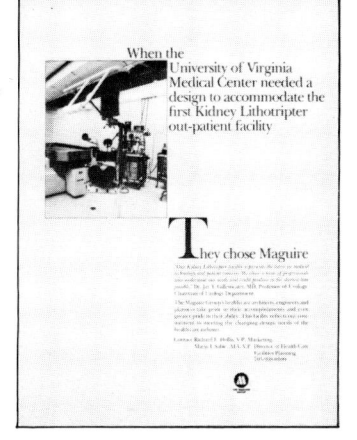

Maguire Group

This architectural and engineering firm specializes in health care planning. Its advertisements use bold black type against a white background and a four-color photograph of a project. The ads are simple and sharp; text is short and to the point. Many of the ads feature a quote from a client. All the ads include the firm's address and prominent logo.

Direct Mail 63

Gilbert/Commonwealth

This engineering consulting firm created a corporate ad campaign geared specifically toward the utility market. It was considered a "puzzle" campaign, and was designed for engineers who solve puzzle-like problems. For the campaign, puzzles were placed in trade magazines. The ad told readers that they could write to the company for the answer.

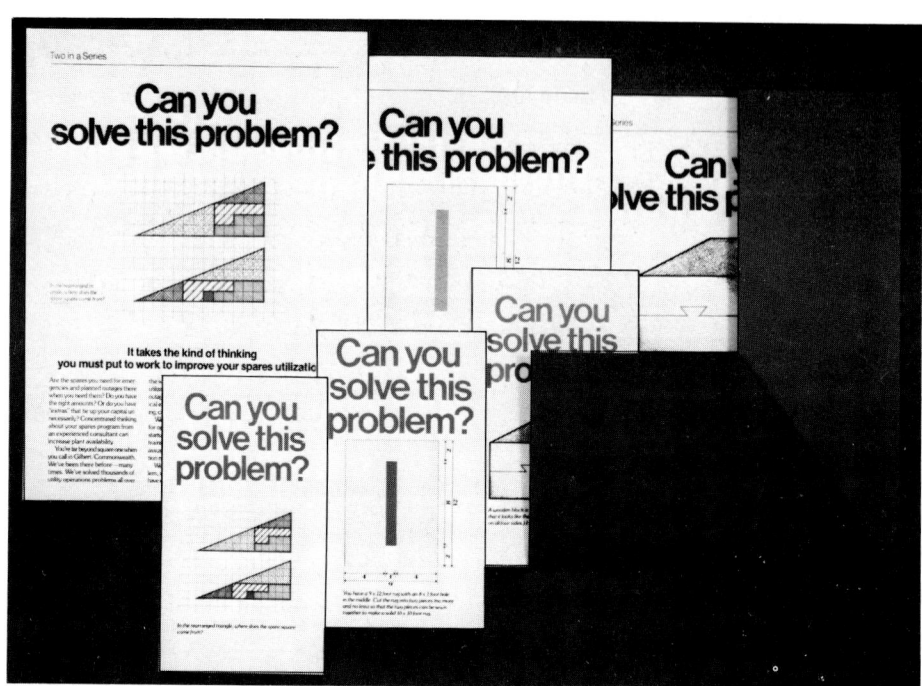

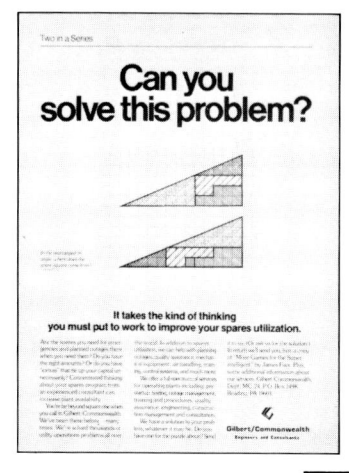

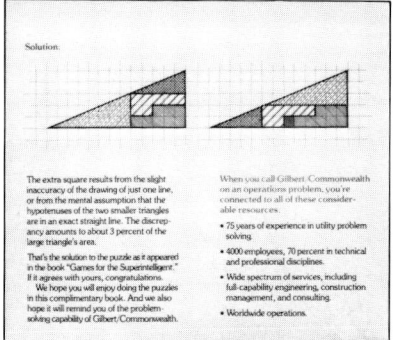

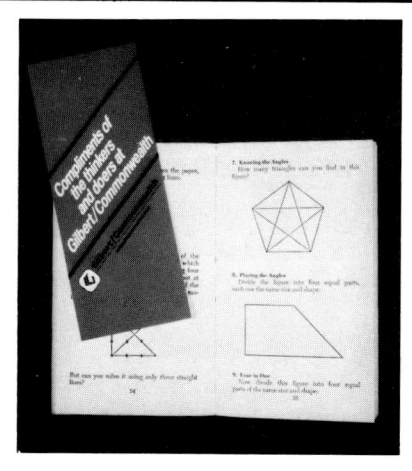

As a response, the firm sent the reader a complimentary book, from which the puzzle was taken; a bookmark; and a tag card with the correct answer. The folder was part of a follow-up campaign, and included information about the firm. The text in the ads is printed in black, red, and white. All material except the book carries the firm's logo and name.

SHWC

An innovative public relations program was initiated through an off-beat poster campaign. The posters were directed towards developer clients and used a play on the local phrase "Dealmaker." The posters were so popular that the firm developed an annual "Dealmaker Award" program to honor those people who have most effectively improved communications in Texas through significant real estate transactions.

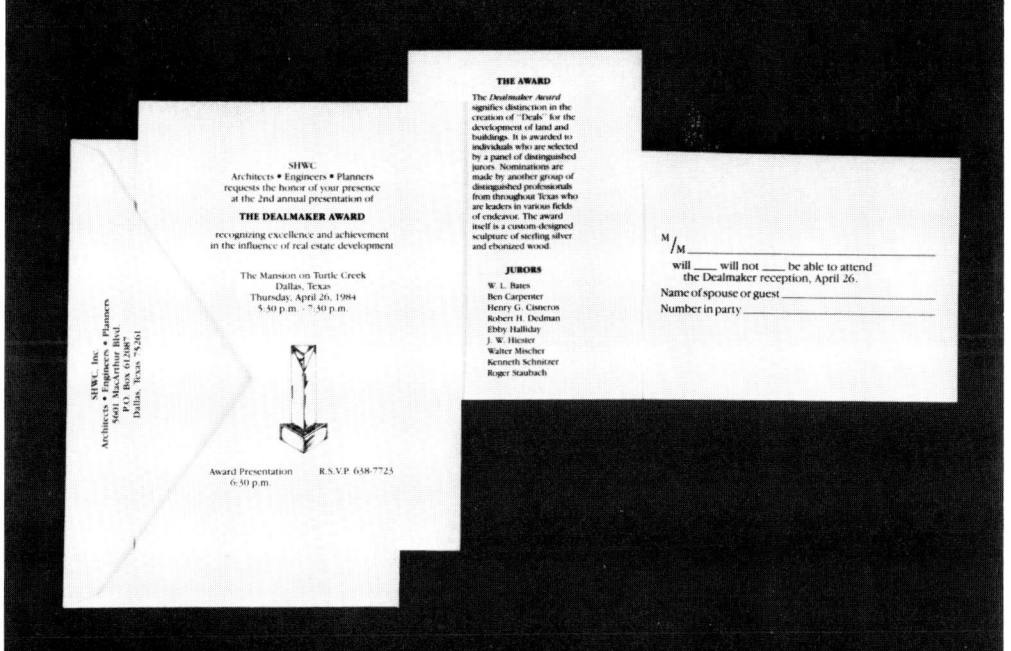

This A/E firm launched a public relations campaign to create opportunities to get to know the leaders in real estate development with whom direct contact might otherwise have been very difficult. An annual "Dealmaker" awards ceremony takes place to honor Texas' top dealmakers in real estate development. The many printed pieces created for this yearly ceremony are: invitations, posters, announcement cards, nomination ballots and the actual ceremony program. All pieces (except for the full-color poster) are printed almost entirely in black and white. The logo is shown in red.

Direct Mail 65

Honeywell

The pop-up advertisement used by this manufacturer shows the potential client how well equipped it is to handle their needs. An opening page of white type on a black background poses a question that clients would ask. The centerfold spread then answers that client's question with a scale replica of their expansive facilities and a point-by-point breakdown of its capabilities. Turning the page, there is an actual photograph taken from the tall tower overlooking the facility. The text further discusses the firm's capabilities and gives a number to contact.

Transamerica

This magazine advertisment was inserted into several different magazine titles. The first page states: "Would the most innovative insurance company in America please stand up," which leads into the pop-up of an over-sized Transamerica building and the surrounding city of San Francisco.

David Paul Helpern

These Christmas cards are modelled after buildings designed or restored by this multi-disciplined firm. The cards are three-dimensional freestanding pieces, about a foot in height or width (depending on whether the project is a high-rise or a low-rise building). The firm's name, the date, and a simple salutation run across either the base or top of the card. The gymnasium card also reveals the inside of the gym: a basketball court complete with spectators, cheerleaders, and players.

Gilbert/Commonwealth

A pop-up was created by this firm to draw attention to a problem faced by many of their potential clients. The three-fold piece fits into a standard envelope for mailing. The message unfolds with the piece and when completely unfolded, a utility pole pops up and the full message is given along with whom to contact. Follow-up calls were made after this piece was mailed.

Direct Mail 67

Morris/Aubry

Full-page advertisements for this firm feature architects standing with scale models of their projects. Quotes from the architects are used as well as a brief statement about the company. The ads give a feeling of confidence in design and, in essence, that the architects stand behind their projects.

HLW

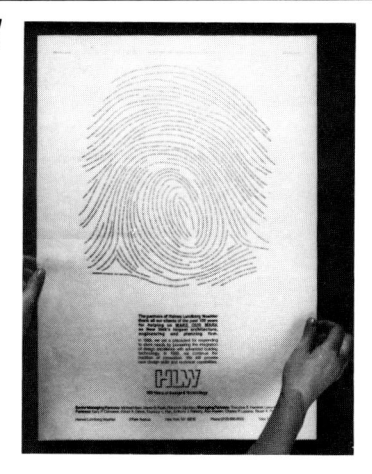

One element of this A/E firm's centennial program included an advertisement thanking many of its clients. The advertisement was placed in The New York Times and appeared as a full-page in the business section. The client's names are placed in a pattern to resemble a fingerprint and the copy for the ad thanks them for helping the firm "make its mark" as New York's largest A/E and planning firm.

THE NEW YORK TIMES, THURSDAY, MARCH 14, 1985

The partners of Haines Lundberg Waehler thank all our clients of the past 100 years for helping us <u>MAKE OUR MARK</u> as New York's largest architecture, engineering and planning firm.

In 1885, we set a precedent for responding to client needs by pioneering the integration of design excellence with advanced building technology. In 1985, we continue the tradition of innovation. We still provide new design skills and technical capabilities.

HLW

100 Years of Design & Technology

Senior Managing Partners: Michael Maas, Martin D. Raab, Robert A. Djerejian; **Managing Partners:** Theodore S. Hammer, Leevi Kiil; **Partners:** Gary P. Cirincione, Edwin A. Dirkes, Seymour L. Fish, Anthony J. Flaherty, Alan Kaplan, Charles P. Lazarou, Stuart K. Pertz.

Haines Lundberg Waehler   2 Park Avenue   New York, NY 10016   Phone (212) 696-8500   Telex 127720

40

SPECIAL EVENTS: 4

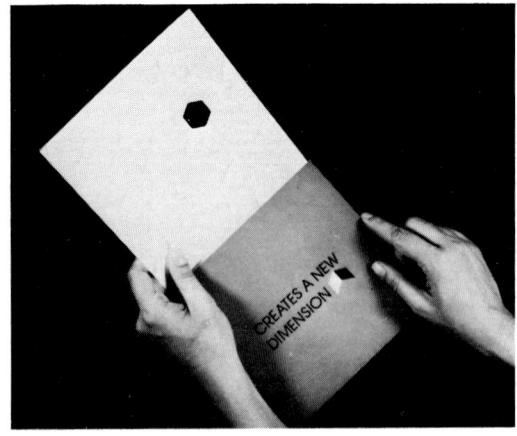

Burt Hill Kosar Rittelmann

This special events piece was created especially to announce the merger of two firms. The background is gray with a white, gray and black hexagonal design and a line of copy. When the piece is opened, the reverse of the cover is white with a die-cut hexagon; the facing page is gray, with another slogan, and a part of the hexagon appear. When the piece is opened again, creating a vertical spread, the entire piece appears with black type and the name and logo of the two firms is revealed.

Holt + Fatter

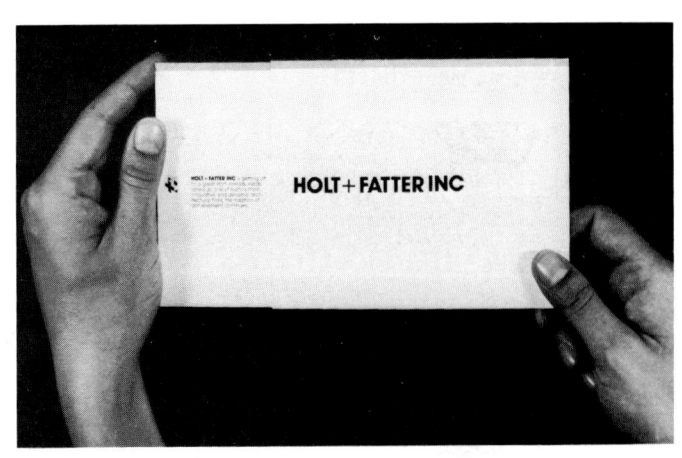

This mailer piece was designed to be an announcement of a recent merger. The original name of the firm is intact when first viewed, then changed to the new name when the inserted panel is pulled out. A brief explanation is also given.

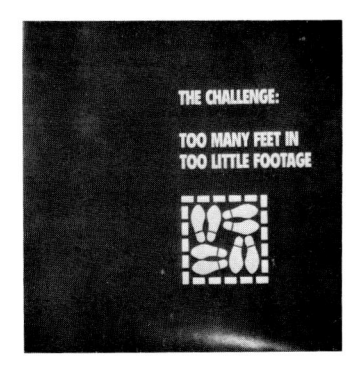

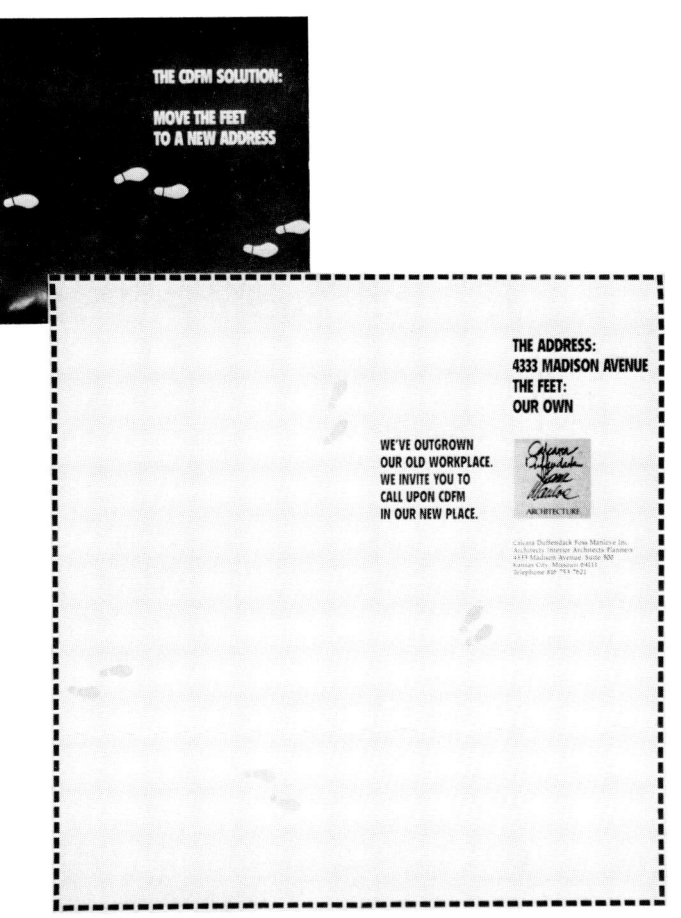

CDFM

This piece announces the firm's new location. On the cover, the copy and illustration complement each other. The gray-and-white color scheme gives the piece a contemporary look. The first unfolding reveals a solution to the problem which was presented on the cover: "Too many feet in too little footage." When the piece is finally opened all the way, the tones are reversed; now the type is gray on white. The visual is enlarged to indicate the enlarged working space, and the firm's logo appears again.

ISD

This interior space design firm needed to send out an announcement that it had changed its location, but kept the same phone number. A 4" x 4" inch card was sent, which folded out into a four-panel piece measuring 4" x 16" inches. The first panel has the firm's logo/monogram and the next has its old address. The following panel has the firm's phone number and the last panel has its new address.

Special Events 73

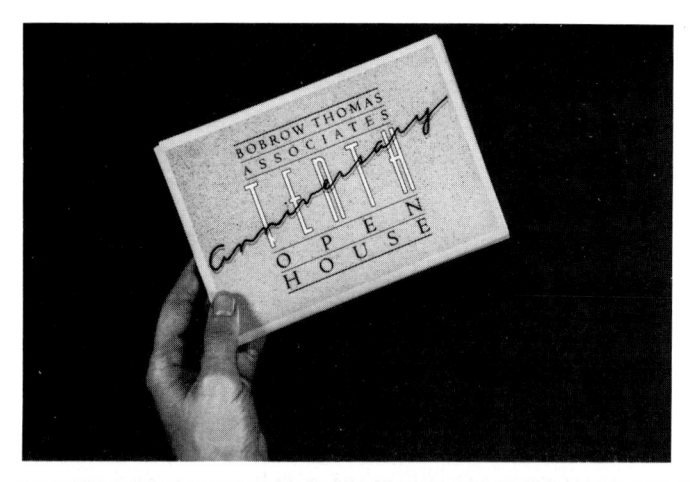

Bobrow Thomas

The front cover of this design firm's tenth anniversary open house invitation uses three type styles in harmony. The piece consists of twelve accordion-folded panels. Each panel reveals a group shot of the firm's employees in front of a blackboard. In each of these photographs there is a new scribble on the blackboard. The black-and-white photographs are printed with a mezzotint screen and a beige color tint. Each person is listed by name on the back of each panel. The piece is playful and gives the impression of a school yearbook. The back cover lists the time and place of the open house.

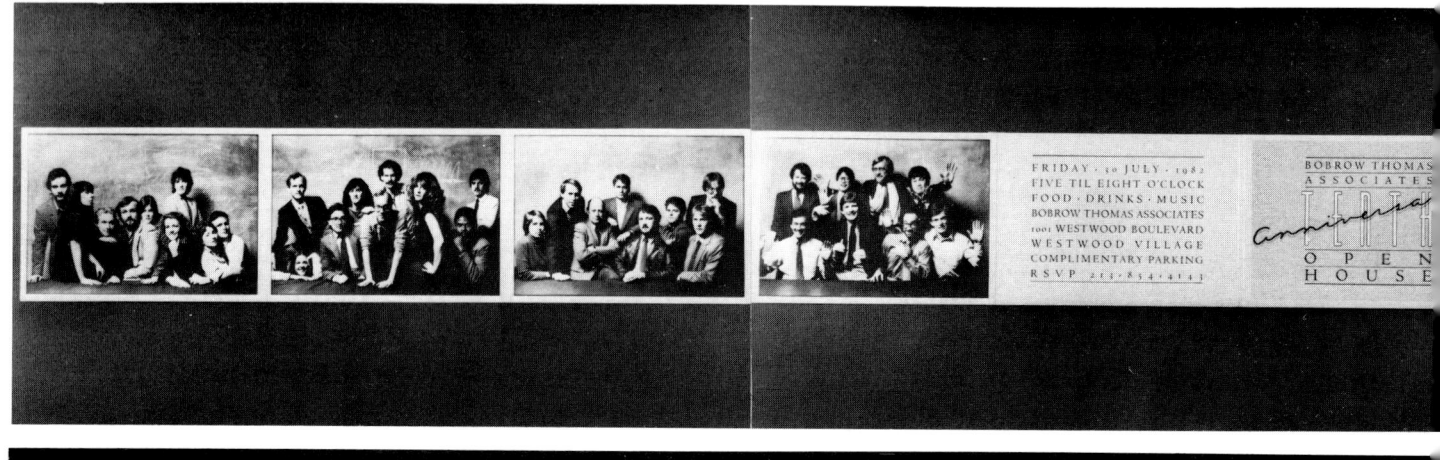

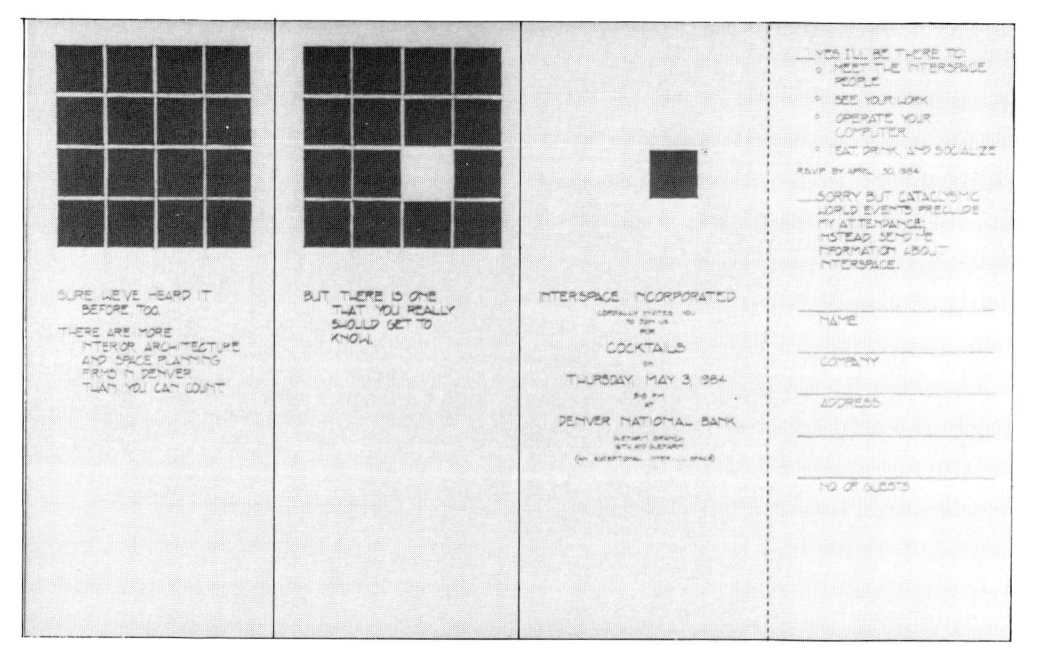

Interspace

This design firm created an invitation for potential clients to investigate its work and offices and operate their computer. The piece is pale gray with a slate gray for accent. The front cover shows a grid, and the type appears to be handwritten, in the same slate gray. The invitation folds out (like an accordion). The second panel again shows the grid, with more type; the third panel shows one block from the grid and more type. The visual is directly linked with the copy. The last page (with a perforated edge) is the mailer.

LZT

For its special events piece announcing the opening of a new office, this architectural firm chose to have a ribbon cutting ceremony, by each of its readers. The black, white and silver piece is designed as a gatefold. The city skyline was used with a full moon backdrop. The activity of cutting the red ribbon to discover the contents was decided on. Inside, the announcement and the skyline graphic are continued. The accompanying envelope shows the city skyline and the firm's address. A small pair of scissors was sent along with this piece.

Special Events 75

Greenboam & Casey

A small accordion-folded calendar was sent out by this firm to its clients and prospective clients. The piece uses yellow on the first panel and subtly graduates to orange then to red. The first letter in the name of each month was used to represent that month and a different type face was chosen for each letter.

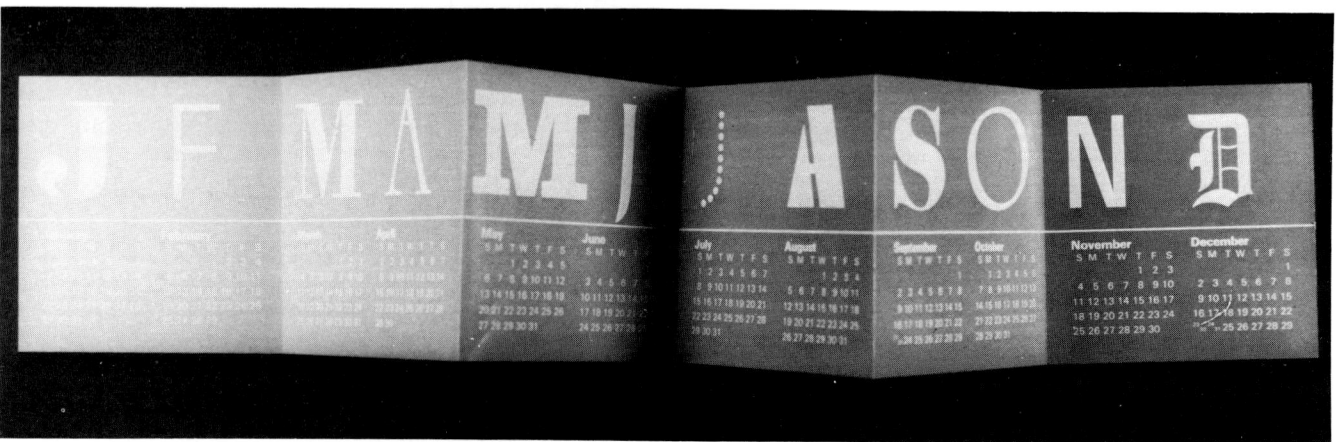

WTW

This architectural and planning firm created this calendar showcasing a large university project that it was involved in. Full-color photographs showing different views of the project are used opposite the calendar on every page.

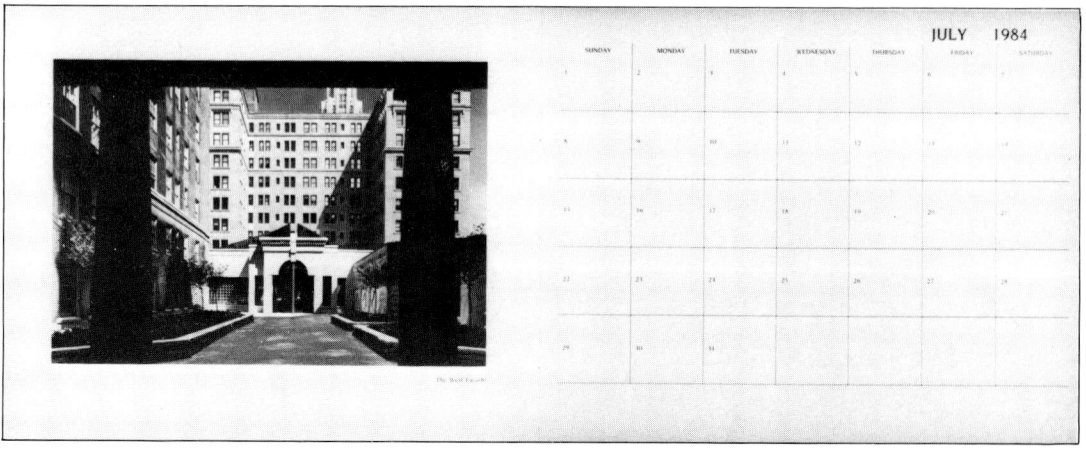

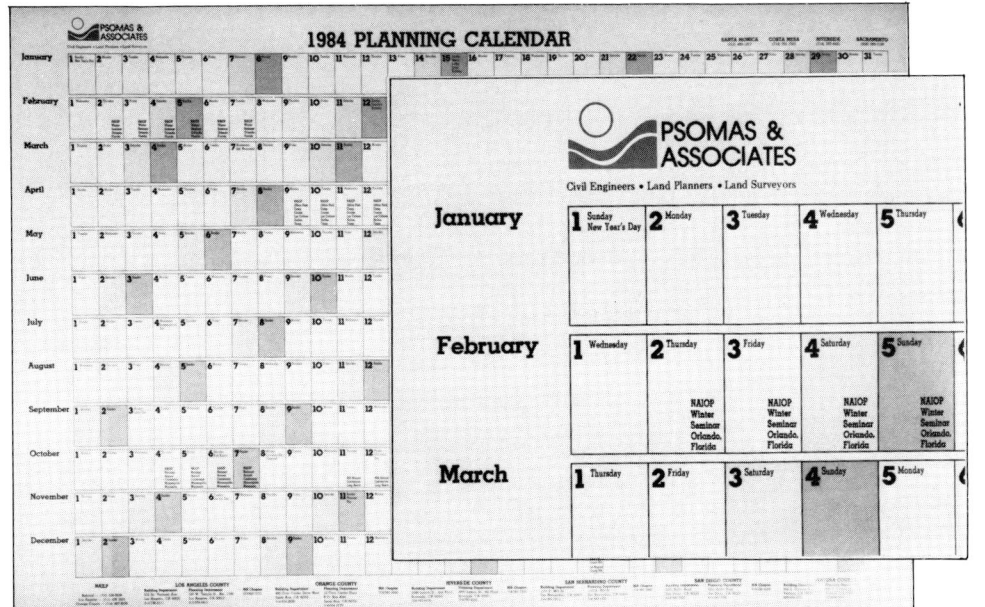

Psomas

A functional piece was created for this engineering and planning firm to remind people of the firm. The yearly calendar lists all recognized holidays, and provides a schedule of several major conferences and seminars around the world relating to their field of expertise. The piece is black, white, and gray, and features the firm's name and logo.

The Callison Partnership

The Christmas calendar was developed to function as a work of art as well as a marketing tool, and to be hung on a wall as a reminder of the firm. A tissue overlay was personalized by handwritten signatures of staff members. Special clients were sent framed calendars.

SHWC

This architectural firm has designed a calendar that measures 4" x 36½" inches. The calendar was mailed in a red mailing tube which sets it apart from most other calendars which are usually sent in flat envelopes. The dates and months are printed in bold black and white letters and numerals throughout this piece.

Special Events 77

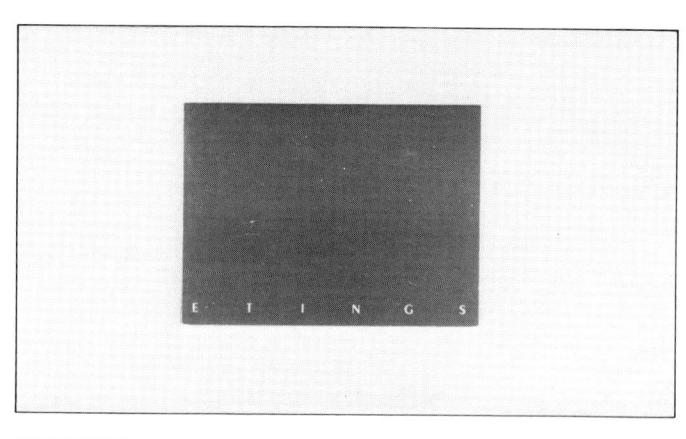

Oliver Design Group

This interior design firm chose a glossy red stock for its Christmas card. The piece has "Season's Greetings" in white on the outside. Inside, a full horizontal spread reveals all the firm's staff sketched in black-and-white caricatures. The company's logo appears at the bottom of the card. The only writing that appears on the inside of the card is the signatures of the people.

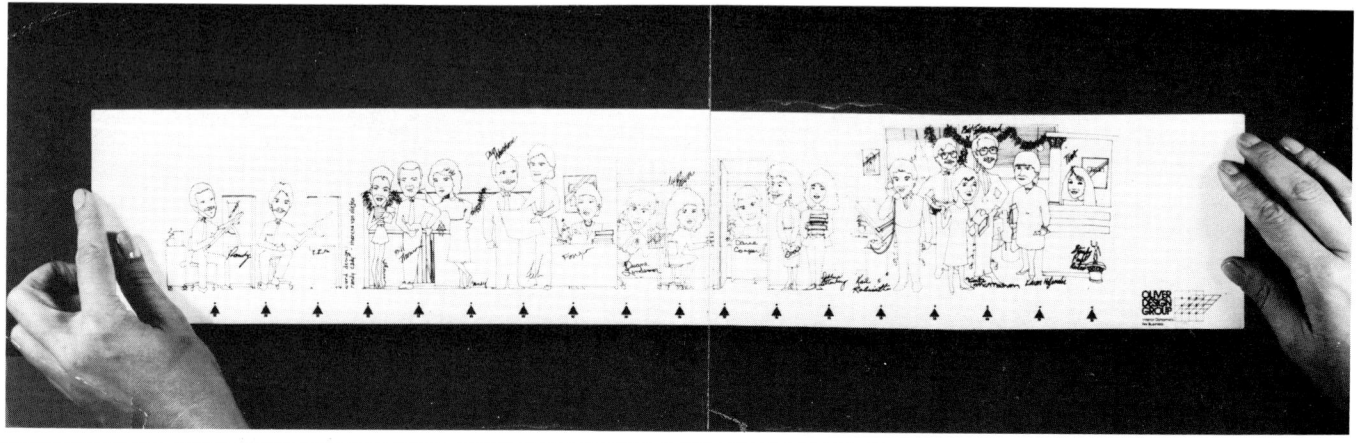

Edwards and Kelcey

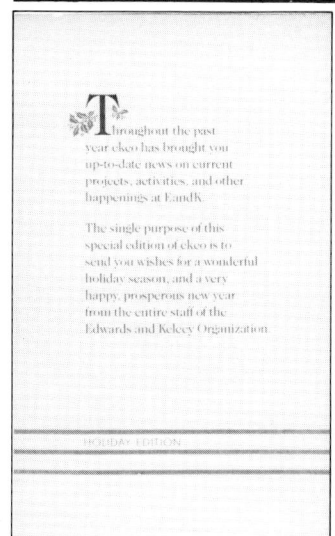

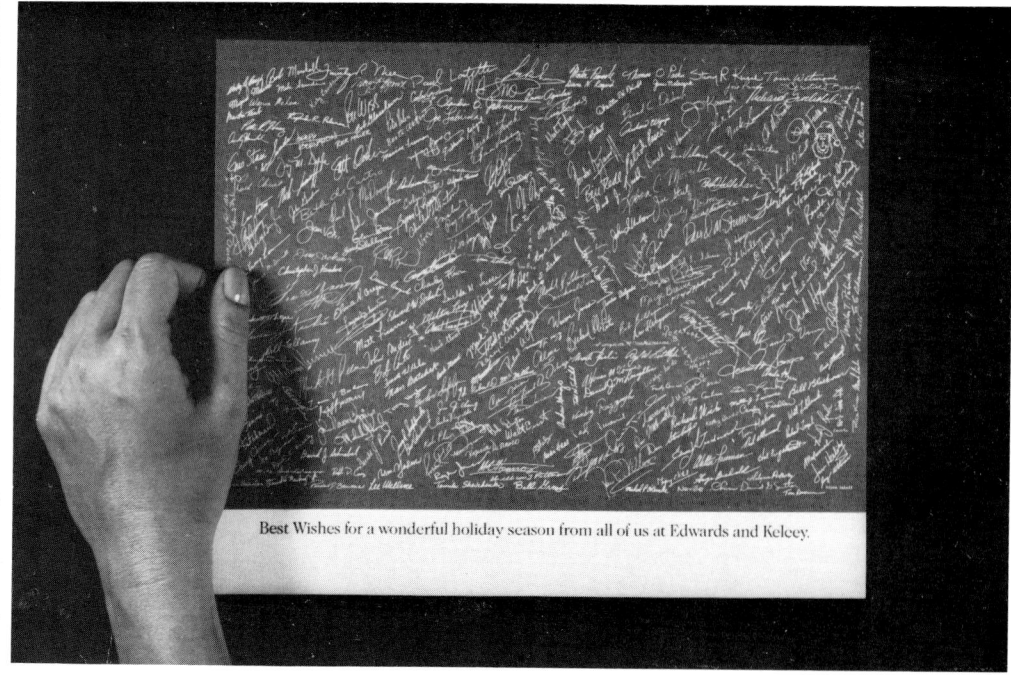

This engineering firm chose a special issue of its monthly newsletter for its Christmas card. This piece is unique in that handwritten signatures of all the company's employees appear in one panel.

RTKL

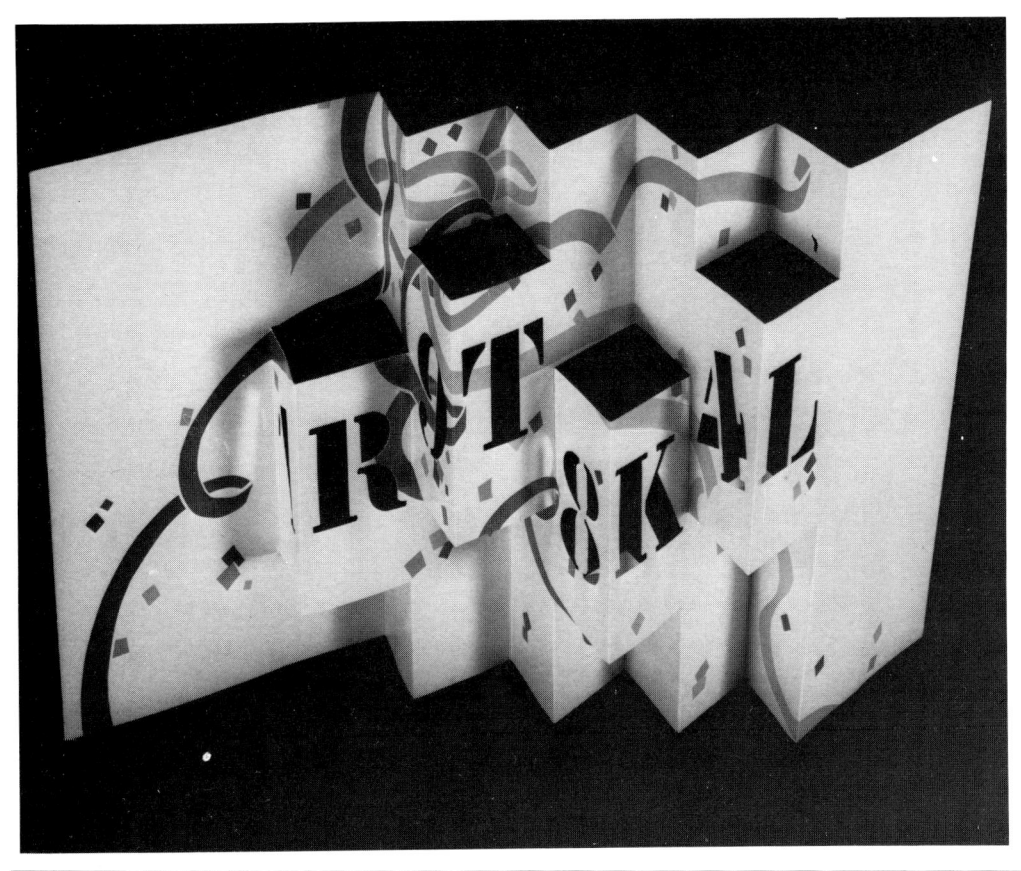

This architectural firm has created unique holiday cards that are eye-catching and convey the desired festive spirit. The 1984 Christmas/New Years Eve card is an accordian-fold die-cut piece with brightly colored graphics cascading over the front. The firm's name and the year are die-cut on top of the pop-out portions of the card.

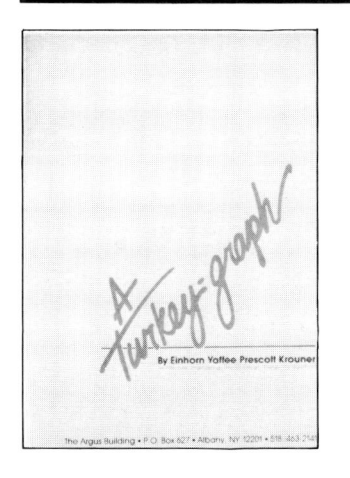

Einhorn Yaffee Prescott Krouner

As an alternative to a standard Christmas card, this architectural firm created a "Turkey-graph" card to be sent out in November. The firm didn't want its card to get lost amongst the usual barrage of Christmas cards that are sent out every December. The graphics in the interior of the card show a graph with an indicator line that resembles the profile of a turkey. The line scales upward indicating a growing number of employees and the fact that the firm has been in business for 10 years.

Special Events 79

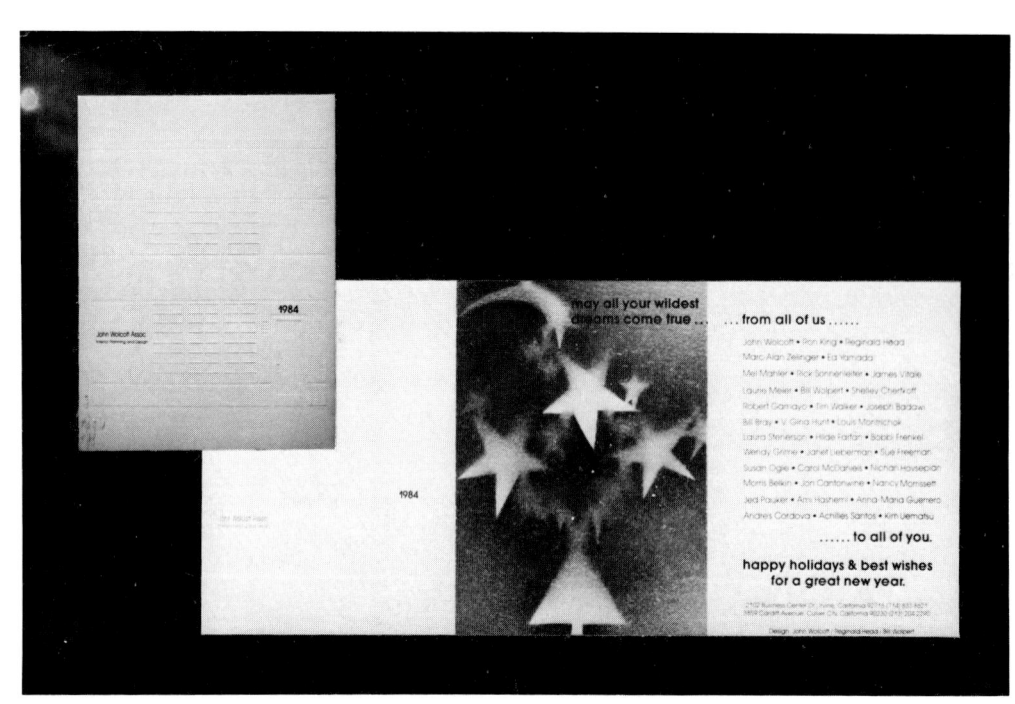

John Wolcott

The front of this greeting card shows a building embossed in white on white. The firm's name, discipline, and the year are shown in blue. The card opens up to three panels: the cover, the middle section (a blue and white visual with stars and a greeting), and the last section, which lists the firm's name and address.

NBBJ Group

This architectural firm chose a foldout Christmas card. The colors are purple, yellow and blue. A wreath is shown on a window, which seems to be softly lit by a candle from within. Inside, there is a blue-and-white snowy visual. A simple, straightforward greeting is included.

Williams Trebilcock Whitehead

For their twenty-fifth anniversary Christmas party, this architectural firm chose a fold-out invitation. The front cover shows the firm's initials in black against a silver background. Wreaths and other Christmas-related visuals are shown. When the piece is completely opened, the numbers 5, 10, 15, 20 and 25 indicate how many years the company has been in operation, as well as pose as numbers of a building. At the bottom of the piece, the company's name and services are listed.

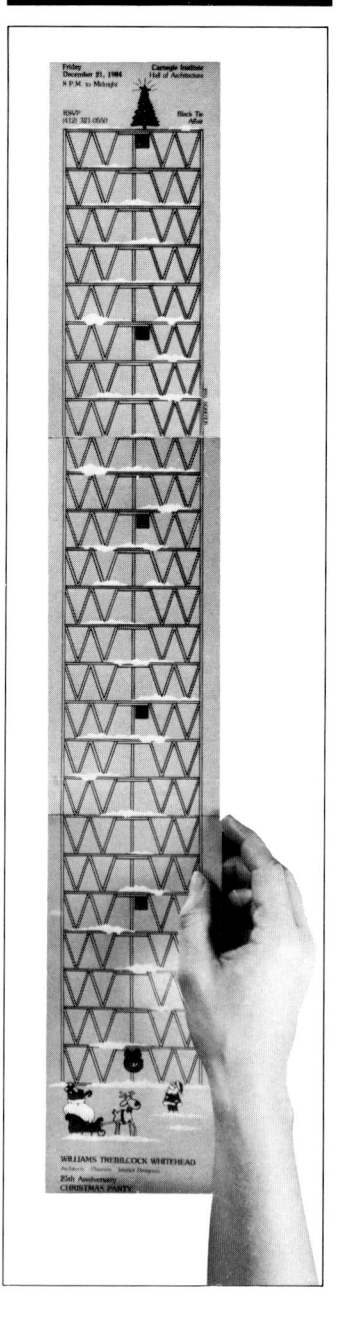

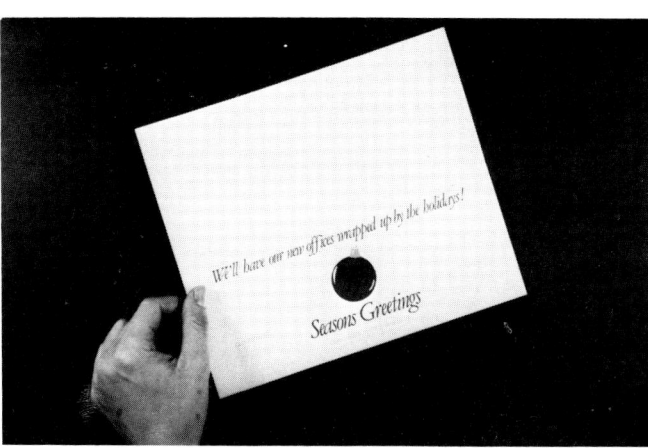

Clark Tribble Harris & Li

For its Christmas card, this architectural firm chose to show a building wrapped up in a big red bow. Inside, the message is festive, and an announcement of its new office location is included. The date of the move is printed in black inside a Christmas tree ball. The old address, as well as its new address, is listed inside.

Preiss Breismeister

For its Christmas card, this architectural firm chose to adorn architectural drawings with traditional Christmas images. The images shown are examples of the firm's renovation work.

Special Events: Christmas Cards

Thompson Ventulett Stainback

This brochure is geared for the same audience as the company brochure, and also for VIPs who influence architect selection and the business media. The fifteenth anniversary brochure shows the firm's key projects over the years, its awards, and every client the firm has worked with. Initially designed to be used with an invitation to the anniversary celebration, the piece could also be used alone. Other coordinated graphic materials used were employee tags, buttons, balloons, and exhibits.

NBBJ Group

A fortieth-anniversary company brochure and accompanying materials were designed by this firm. A four-color visual appears on the cover of this brochure. Two sections of the photograph are cut away and "floated" around it, turning the pieces into graphic bars for the text to fall under.

This cutting away of sections of photographs and using the pieces elsewhere on the page is repeated throughout this piece. A thin rule line is constant at the top of each page, and the text discusses projects and clients. The number forty is repeated on all of the pages, in different typestyles and positions.

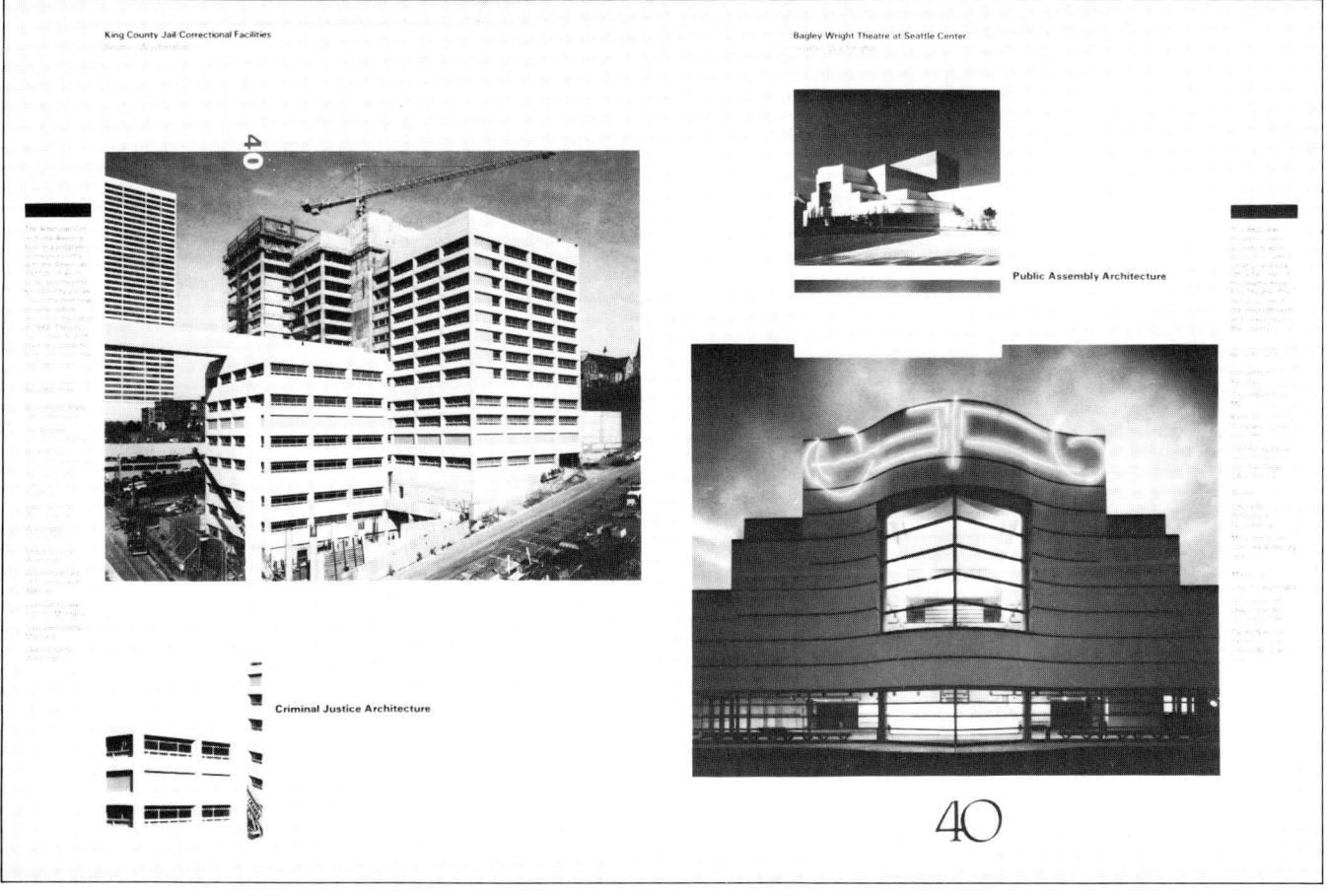

Invitation

A frosted vellum three-panel foldout invitation designed in pale shades of green and yellow allows the reader to see through the piece. The front cover features a graphic and a blank tab rule at the top like that of the brochure. The parking pass and an R.S.V.P. note, both in the same pale colors as the invitation, are included.

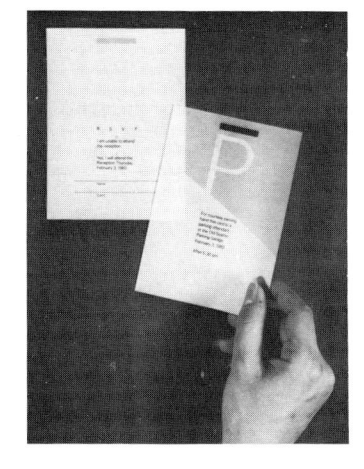

Sikes Jennings Kelly

An art poster depicting one of the firm's first projects was designed to be used for direct mail purposes. A dramatic color photograph of the facade is framed in a slate-gray border. The copy reads, "Architecture Reflects Image." At the bottom of the visual, the copy is reflected in a mirror image. The type is light beige and sky blue. On the back of the poster, the copy reads, "And Good Images Don't Get Built in a Day." Additional text discusses the firm in five short paragraphs. Details of the cover project are also given, as well as the firm's address and telephone number.

Murphy Group

A poster was designed as an invitation to a St. Patrick's Day party. The design was chosen after many tools of the design field were considered. A swatch of material was finally agreed upon. The swatch, with traditional pinking-shears edges and a hole at the top (where all the swatches hang together), was green (for St. Pat's Day), with purple and white as accents. The invitation was designed in a 12x23½-inch glossy finish poster format to attract attention from people in the field. The green type against the white field stands out and provides the reader with all the necessary information.

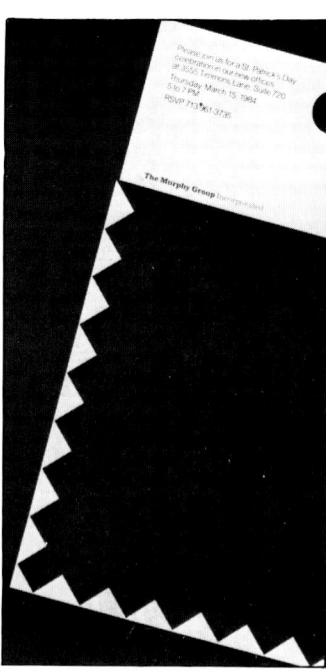

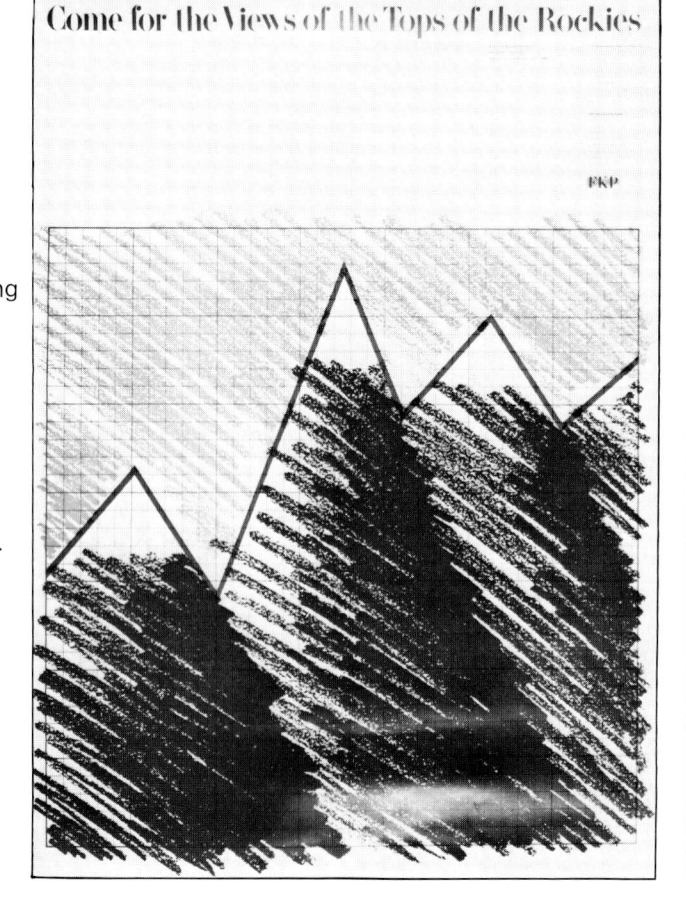

Falick Klein Partnership

This architectural firm designed a poster announcing a health care convention open house. The blue and green graph is designed like those seen on the progress chart of a patient, yet the visual also resembles the Rocky Mountains. The type is small and does not detract from the visual, but gives relevant information. The eye-catching copy is placed against a white field and set off from the visual. The firm's logo appears with a wash of blue similar to the blue sky above the mountains.

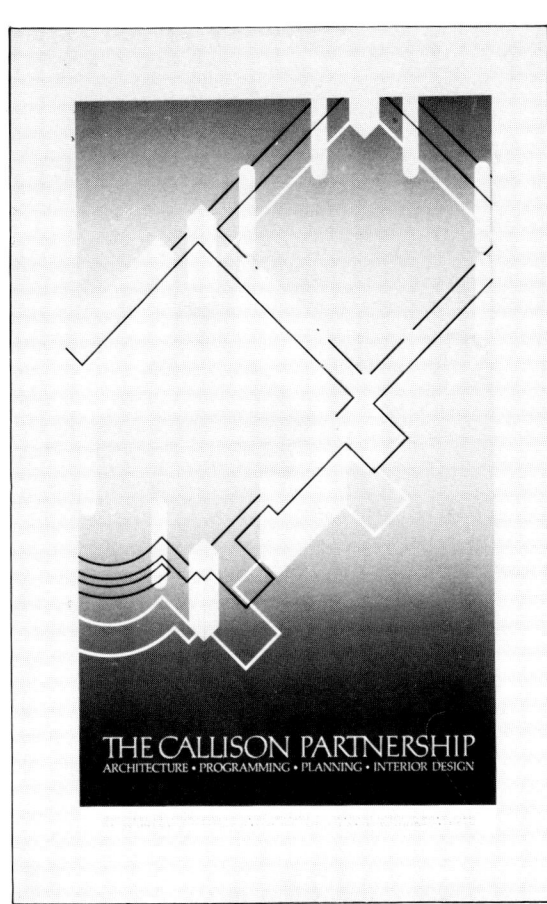

The Callison Partnership

This firm has also designed a poster as part of its marketing materials, which are extensive. The poster displays a line drawing of a building. The building blocks are in yellow and white, the connecting lines in white and maroon. The background shades move from gray (at the top) to maroon (at the bottom). The white type gives the firm's name and addresses as well as its areas of specialty. The visual is particularly interesting because it comes out of the left-hand and top borders into the white frame surrounding the visual.

Perez

This interior and engineering planning group designed a colorful poster for the Louisiana World Exposition. The poster, in cartoon style, shows a medley of activities flags waving, mermaids standing guard, balloons taking off, families watching the events, alligator tamers, clowns, musicians, and boy scouts. To keep with the playful nature of this poster, the drawing overflows in the white border that surrounds it. The bottom portion of the poster has the white type set against a black border.

Special Events 85

Giffels/Harte

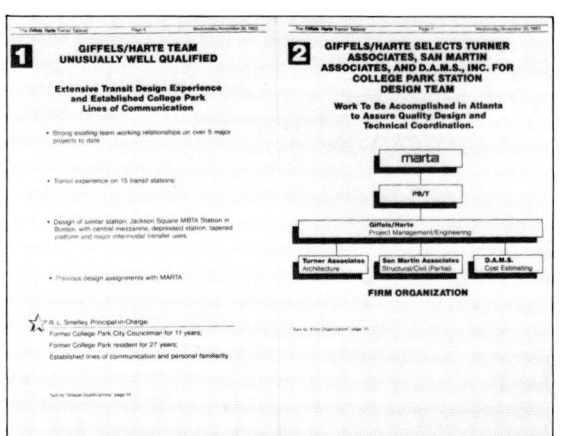

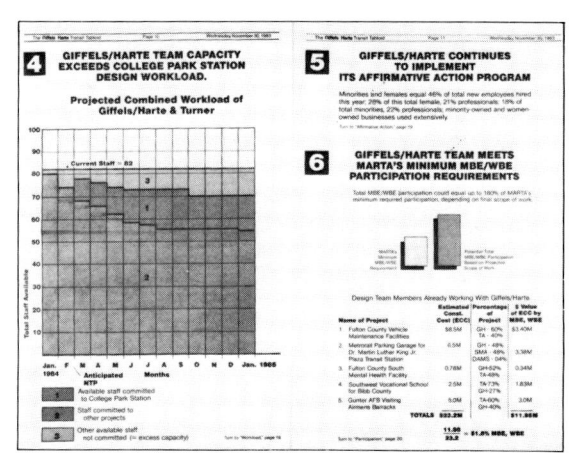

This presentation piece uses a newspaper format. The primary objectives were to present clear, concise and accurate responses to ten specific questions asked by the project's selection committee. This piece was able to answer the committee's questions without the aid of slides or other standard presentation media. A series of interviews was conducted by the firm of local public figures and direct quotes from those interviews were used in the tabloid. A 40" x 60" inch enlargement of the piece was placed at the front of the presentation room and each committee member received their own personal copy. At the conclusion of the presentation yellow subscription cards (which were fashioned after an actual MARTA rail pass) were passed out to the committee members and they were again urged to "subscribe to the Giffels/Harte team."

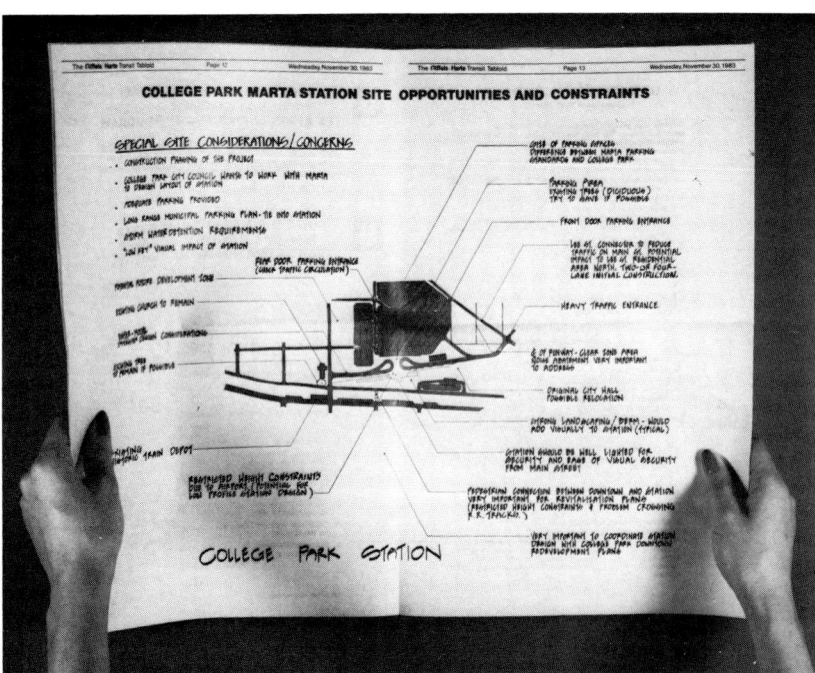

The Grad Partnership

This piece was created to draw attention to the firm's ranking as 44th in the nation and first in its home state of New Jersey. The number 4401 was chosen as the firm's "lucky number" and lottery tickets were sent out using this number. The number is printed on the cover of this three-fold mailer piece. The text on the second panel discusses the odds of winning the lottery and the odds of choosing the wrong design firm. The text then goes on to say that you would be reducing the odds in your favor by choosing a firm with 78 years of experience. When fully opened, the lottery ticket is revealed with information on how to find out if you've won.

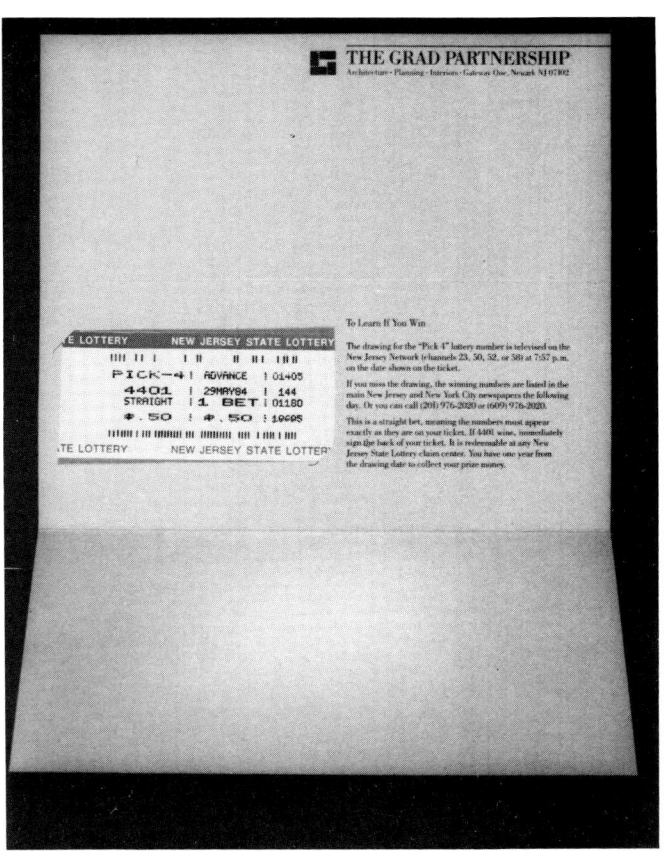

NBBJ Group

This architectural firm needed to create a presentation piece for a proposal to build an engineering and computer science building for Seattle University. The piece that was created is fashioned after the popular board game "Trivial Pursuit." A series of question and answer cards was presented in a small box featuring the name of the campus on the front and the name of the proposed project on the side. Each card stands upright in a slotted holder at the bottom of the box. Five categories of questions are given on each card and the answers are supplied on the back. The categories consist of Seattle University, project facts, technical facilities planning, firm profiles, and scientific facts. The answers, all led to the selection of the NBBJ team.

Special Events 87

NBBJ Group

This multi-disciplined firm needed to create a booth for a trade show in the area of biotechnical laboratory planning. It chose as its visual theme a woodcut print of Pasteur's laboratory. A dividing partition was designed using an enlargement of the print along with two small pamphlets whose covers use sections of the same print. These take-away pieces discuss the firm's capabilities and who to contact for more information.

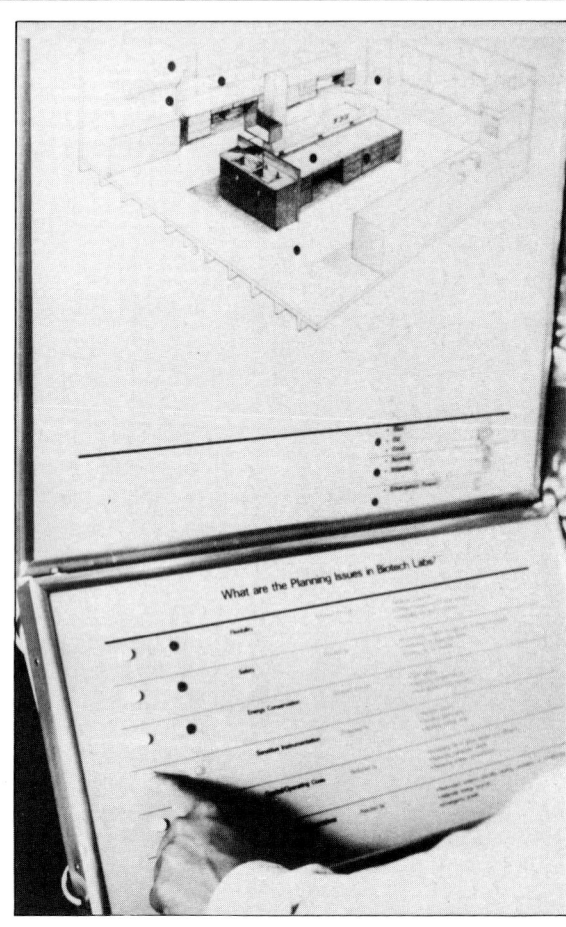

Also provided is a mechanical push-button list of the planning issues involved in biotech labs.

Noah Banking

This computer service firm wanted potential clients to get the hands-on experience of its newest information system. A spacious area was provided equipped with the actual equipment for people to see and try when they entered. Take-away brochures in full color were also provided. The brochures show the system in use and point out its assets.

Special Events: Trade Shows 89

Vanderweil
MORRIS ☆ AUBRY ARCHITECTS
MINATED
MANUSCR
AWBW
BISMARCK, P.C.
ARCHITECTS & PLANNERS
Rudolph/Libbe/Inc.
rchitects
pl
The Creative Process. Design is just one aspect of hospitality project delivery. Index has developed a complete set of services that take a project from a rough sketch to a finished product quickly, efficiently, and with great results.
In the beginning we participate in studies to determine how a project can be positioned for success. Early design concepts—based on budget, scope, schedule, and market conditions—are refined through a series of reviews. Estimates, floor plans, elevations, and sketches showing furniture, finishes, color, and texture are developed. These are the products in our total concept package.
Next, we produce construction documents, including drawings, schedules, and specifications. We help clients prepare and award interior construction and furnishing contracts. During construction, we are contract administrators, site inspectors, quality assurance managers, negotiators—and relentless perfectionists. When an Index project opens its doors we are there, just to make sure everything is right.
Index is prepared to handle the total sequence for hospitality project interiors—from feasibility and concept through construction and start-up. We also offer our services as consultants, or as members of a design team, to provide our type of talent.
CONVENTION CENTER ALBUQUERQUE NEW MEXICO
REPRESENTATIVE PROJECTS
ADAL Computer Building Kirtland AFB, NM
Air Force Housing Cannon AFB, NM
Officers Housing
Albuquerque High School Albuquerque, NM
High School
Allison Cancer Therapy Center Midland, Texas
American Federal Savings and Loan Albuquerque, NM
American Financial Plaza Albuquerque, NM
Anna Kaseman Hospital Albuquerque, NM
Bernalillo County Medical Center Albuquerque, NM
University Medical Center
Biology Building Albuquerque, NM
University of New Mexico
Brashears Center Fullerton, CA
Professional Office Building
Camelback Inn Phoenix, AZ
Resort Hotel
Cancer Research Facility Albuquerque, NM
TOWNHOUSE OFFICE AND HOTEL PHOENIX ARIZONA

COMPANY BROCHURES: 5

Index

One main objective in creating this company brochure was to announce the firm's merger with a national consulting firm. The cover of this piece is mat black with the company's name and logo embossed in the center. At the beginning of the brochure, a quote is used opposite a full-page visual.

ISD

The cover of the folder that encloses this firm's project pages (which vary in length) is red type with a blind embossed logo. The logo is repeated throughout the individual pages to create visual continuity. Enclosed within a square, the logo floats within "interior" space. Inside the brochure, red rule lines emphasize bold type and text. Color photographs of interiors are used extensively. The pages are designed as two-sided project descriptions. The front of the page is color, and the flip side is black-and-white. The back of the brochure includes a partial client list. A back pocket is also provided for the storage of additional material.

Company Brochures: Interior Design 93

PDI

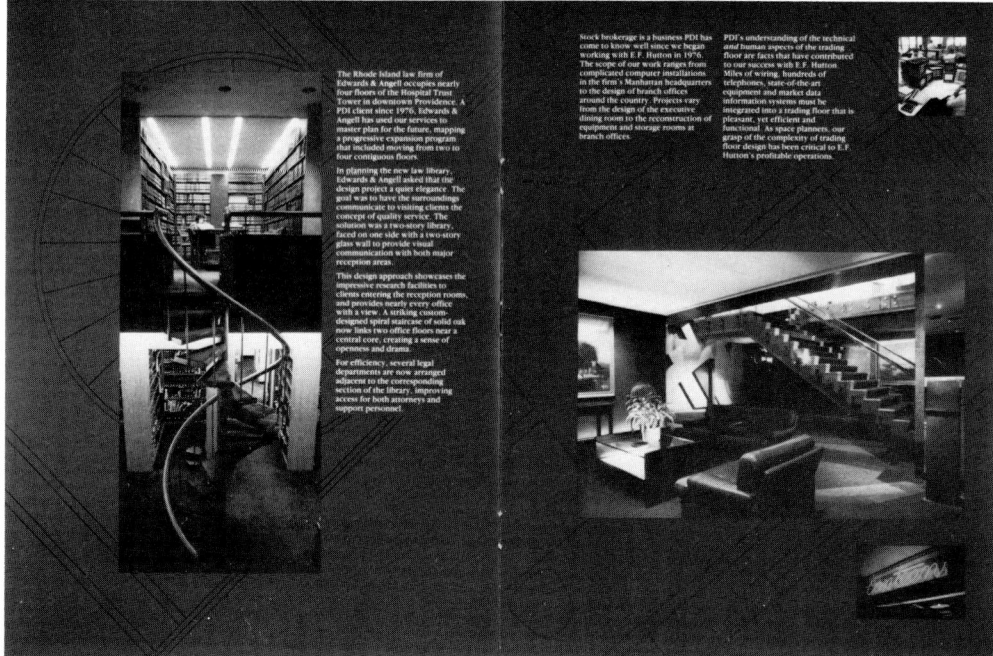

This space planning firm has produced a brochure and project pages using photographs from an audio-visual presentation produced by the firm at the same time. The double use of these visuals worked to reinforce all materials, though all materials can easily stand alone. The brochure uses a black matt background with enlarged portions of floor plans printed in a black gloss spot varnish over the entire background area. These portions of floor plans relate directly to the full-color photographs featured on each page. At the back of the brochure the firm provides a sample flow chart showing phases of production that go into a project. The project pages can be stored in the pocket at the back of the brochure. These pages feature full-color photographs printed on a solid black matt-finish background.

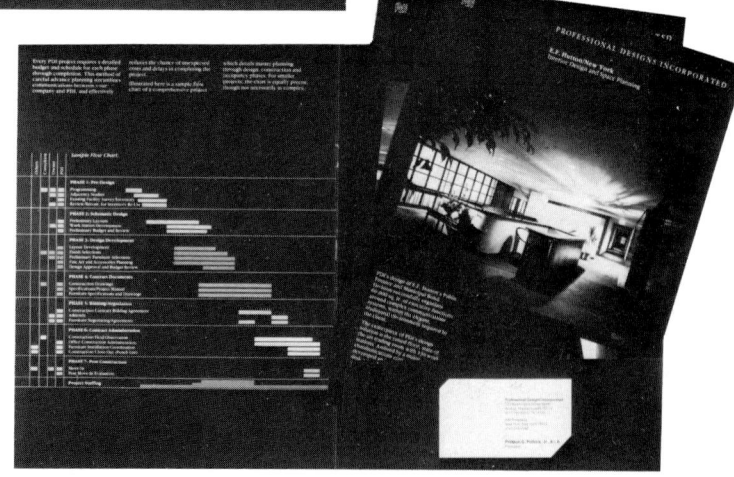

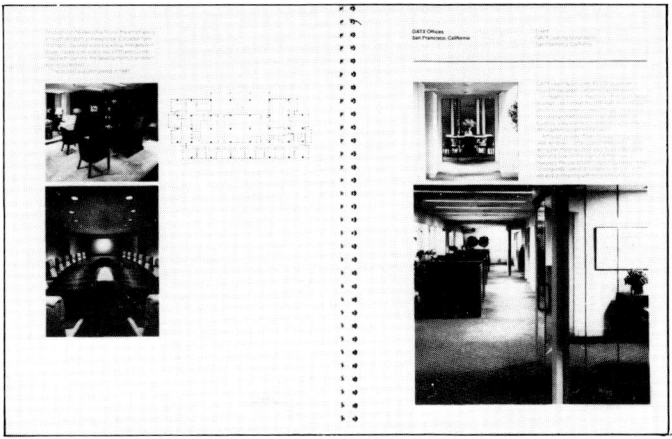

EPR

This design portfolio is comprised of a loose-leaf system which can incorporate project pages. Inside, a spiral-bound brochure is attached, creating a pocket where related material can be stored. Color photographs of the project, descriptive captions, floor plans, and location are all included in the package.

SMS

This firm provides complete architectural and interior design services. Its brochure was designed as a self-contained, comprehensive document aimed at a diversified clientele. The piece consists of project pages using large color photographs, floor plans, black-and-white illustrations, and a description of the site.

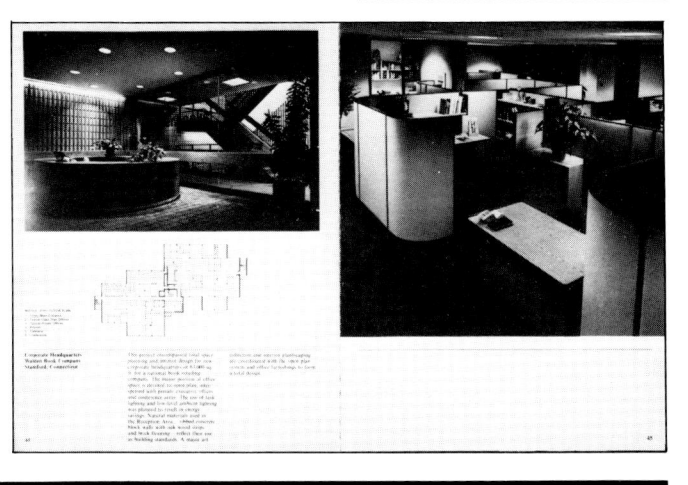

Interior Environments

This architectural and design firm chose, for its company brochure cover, a black matt spot varnish (highlighting the firm's logo) on black gloss with white type. Inside the cover, a plastic spiral binding holds the pages together. Pages are essentially project pages, complete with visuals, descriptions of sites, and headings.

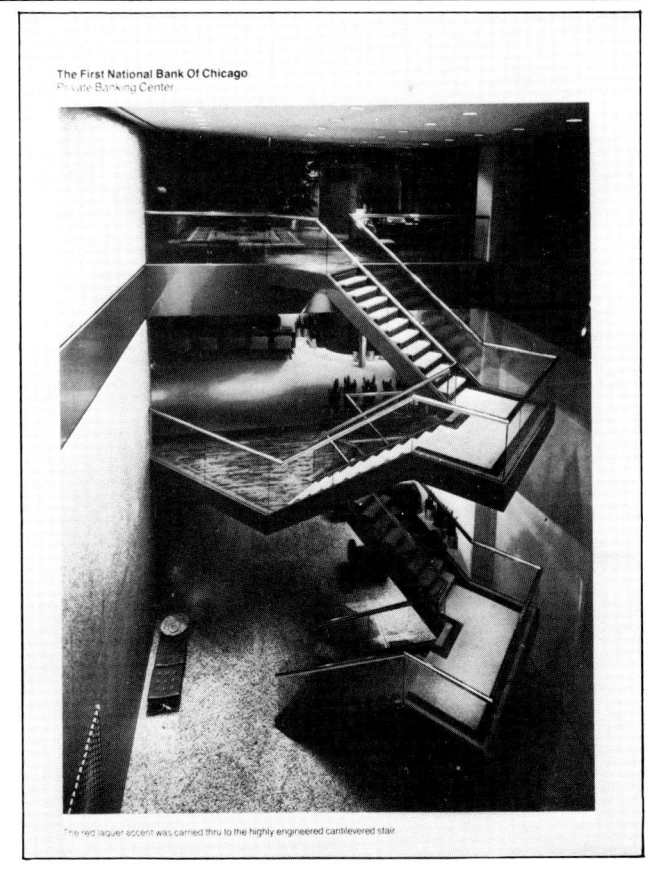

Company Brochures: Interior Design 95

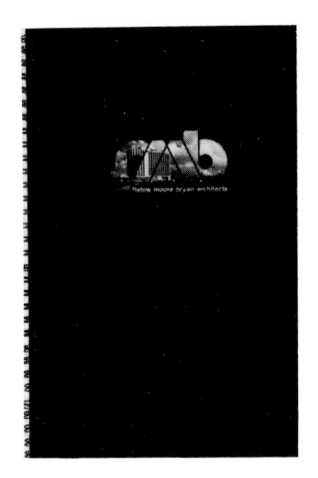

FMB

This narrow metal-ringed spiral-bound brochure features several photographs encased in the company logo, in the brochure as well as on the cover. The brochure's cover and inside pages are all mat black with additional project photos. The type is bold and easy to read.

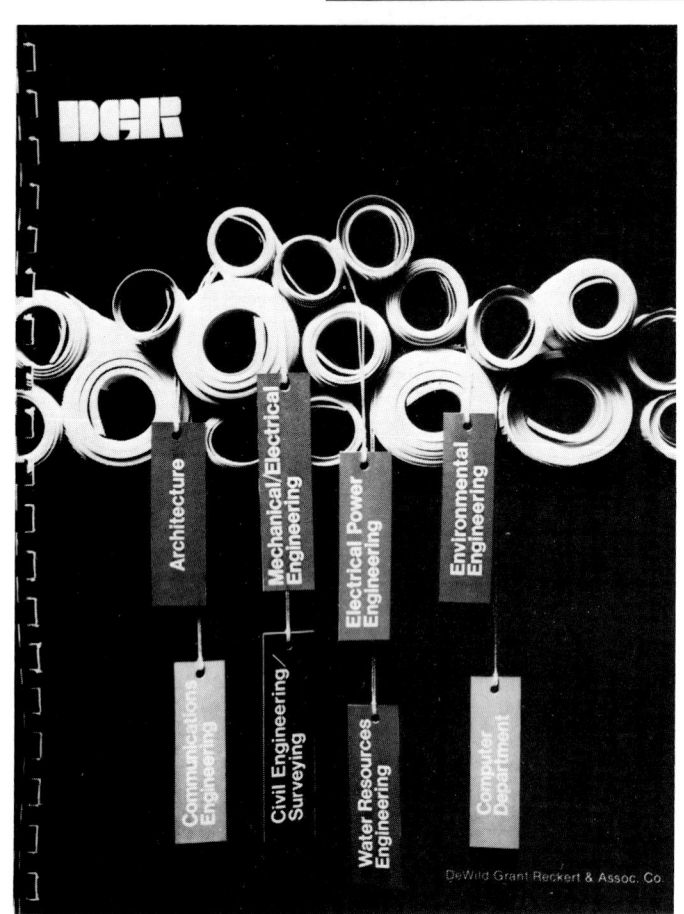

DGR

This firm uses a thick black plastic, continuous-spiral binding on their company brochure. On the cover, hanging tags appear with specialty areas. These tags reappear (individually) at the beginning of each specialty chapter. The piece includes color photographs and text. A list of professional services is given, accentuated by bold white headings.

CDFM

This oversized brochure has been produced in a mat slate gray with a paler gray spiral binding and pale-gray type. Across the front cover is a four-color strip of close-cropped photos, featuring projects and interior furnishings. The inside front cover is a horizontal envelope where additional materials can be kept. The large photos on the inside of this brochure are also bordered, top to bottom, by the same slate gray seen on the front cover. Some pages are bound project pages that show site specifications and renderings.

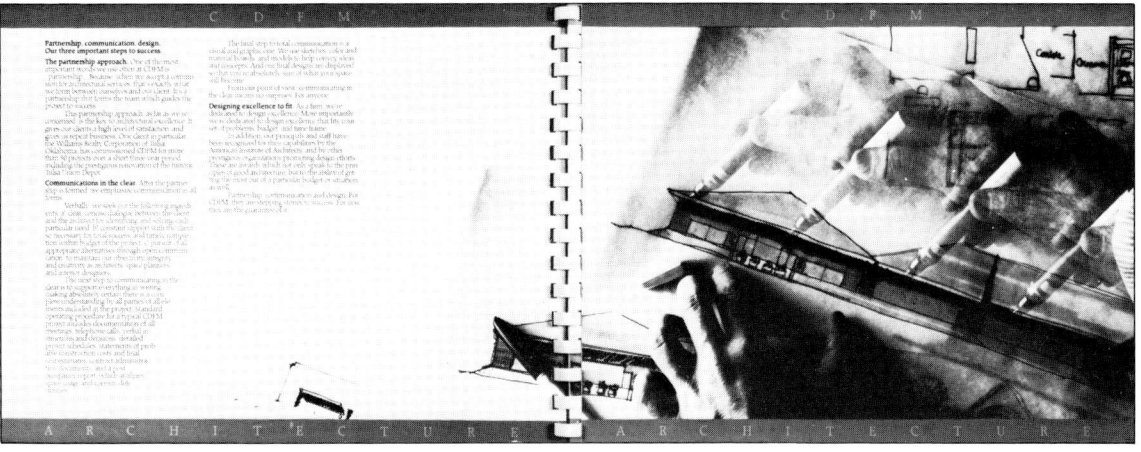

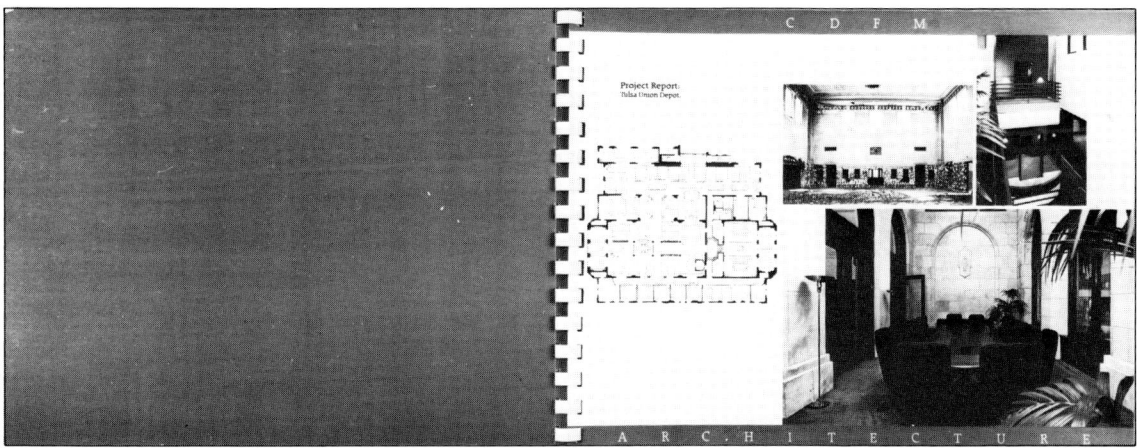

Direct Mail

This mailer has the same cover design as the main brochure. Inside (when the piece is opened completely), the firm's areas of expertise and its philosophy toward architecture are discussed. The firm's services and clients are also listed.

Company Brochure: Architects 97

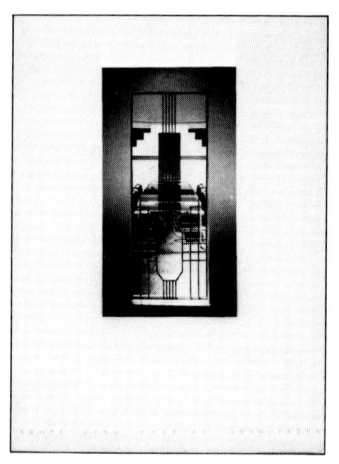

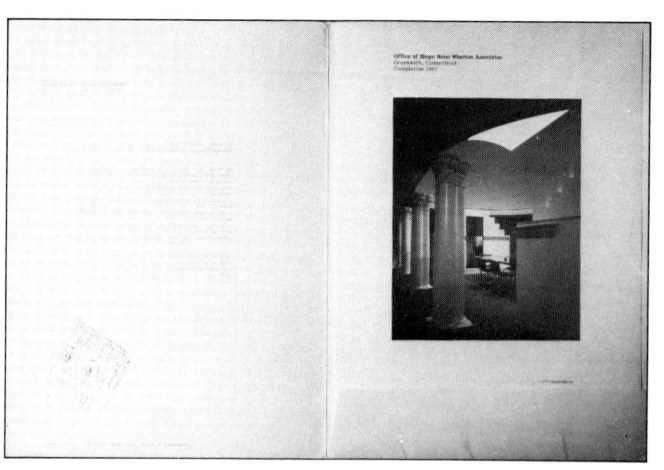

Shope Reno Warton

This firm offers architectural services, specializing in residential and commercial design. A glossy white folder with a centrally placed photograph holds the project pages. Text describing the firm is printed directly onto the inside of the folder's cover. A simple rendering of the firm's office and its address also appear. Project pages are printed on thinner glossy stock. A single color photograph or a grid pattern of four color photographs appears on every page.

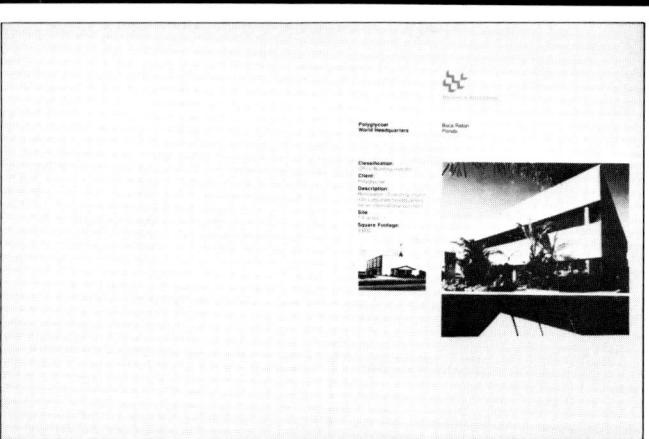

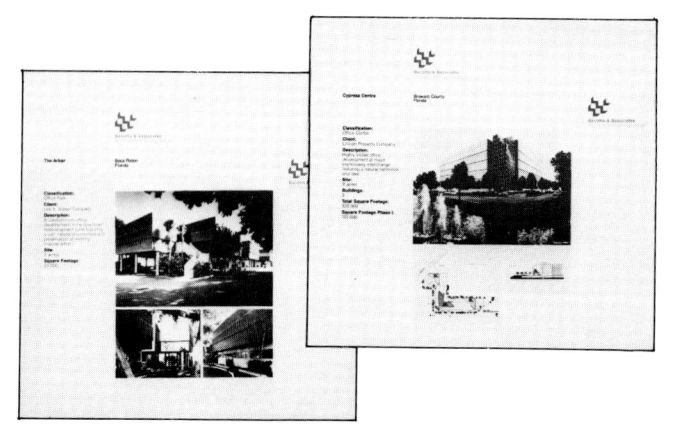

Baretta

This architecture/planning design firm uses a glossy white folder with gray type and design accents. The name of the firm, its specialty, and the company's logo also appear. Inside, the project pages are kept together in a pocket. These pages are all 11" x 11" inches, so they can be cut vertically or horizontally to fit into custom folders. They are printed on heavy stock similar to the folder. Black-and-white photographs of sites, aerial views, and floor plans are all incorporated. The text is minimal, consisting only of the name and location of the specific project.

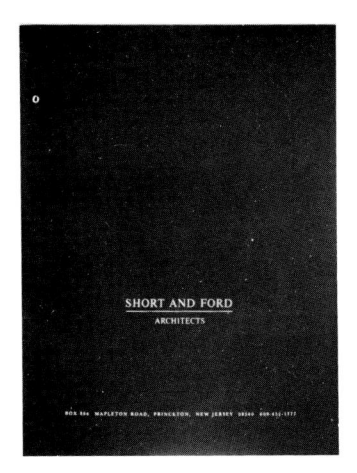

Short And Ford

When this architectural firm recently celebrated its tenth anniversary it created a special brochure to mark the occasion. The glossy white cover, with maroon type, features a four-color photograph of the company's office. The brochure fits into a maroon mat-finish folder. The pages are printed on the same glossy stock as the cover. A maroon rule line is used throughout the brochure. A large, centrally placed color photograph, or a grid formation of four small photographs, is used on every page. Some pages show before-and-after pictures of projects while other show interiors, floor plans and renderings. Several pages are designated as award winning projects.

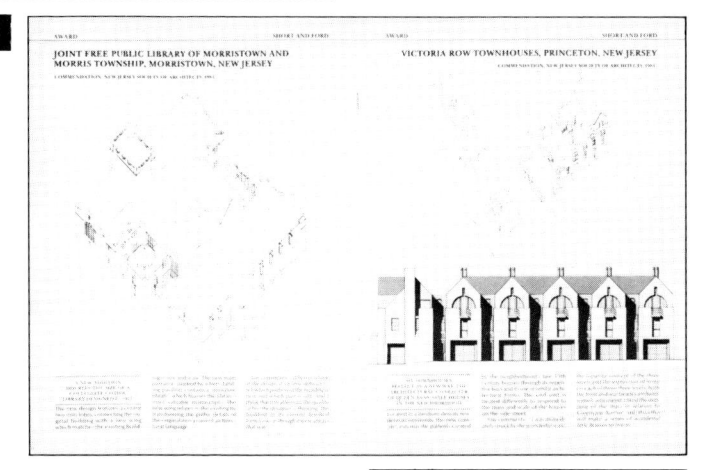

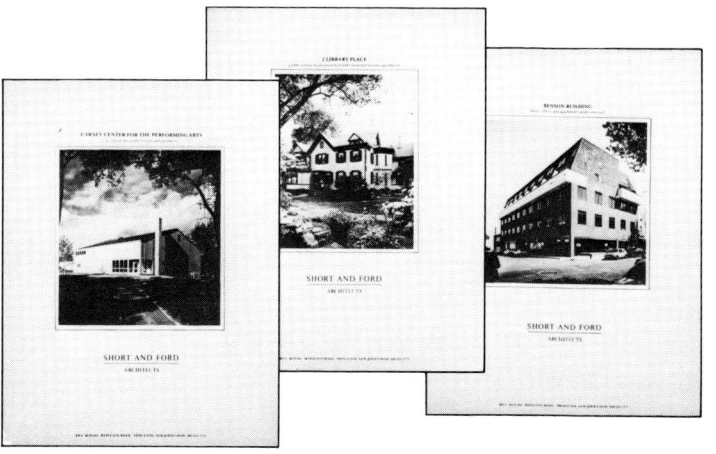

Project Pages

The regular two-sided project pages differ minimally from the brochure's pages. Layout, type, and visuals are much more straightforward, showing the same visuals, but more conservatively than the tenth anniversary brochure. All the project pages include the company name and address.

Company Brochures: Architects 99

Allen-Drever-Lechowski

This architectural firm takes a different approach in the use of its main marketing piece. The cover of the brochure is printed on a dull varnished deep teal blue stock. There are three graphically interesting close-up shots of different projects. The type on the cover, as in the rest of the brochure, is dropped out in white. This brochure is presented as a guide to choosing an architectural firm. Each page includes full-color photographs of projects and the text on each page suggests what different aspects you should look for in an architectural firm, and then explains why this firm meets these standards. A pocket is provided in the back of the brochure for resume pages and a unique questionnaire and matrix on how to evaluate an architectural firm for a project.

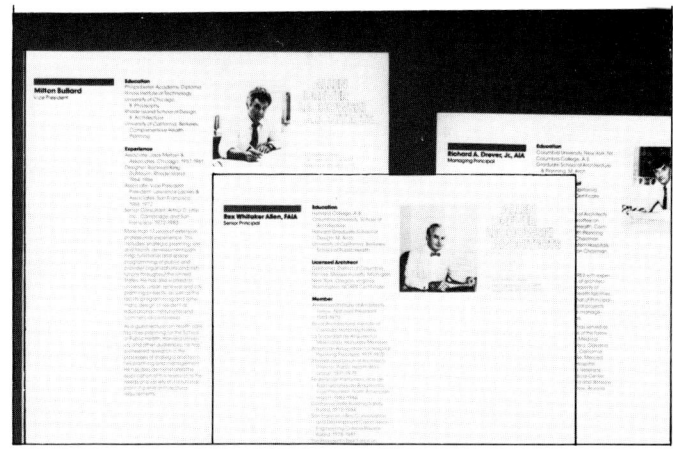

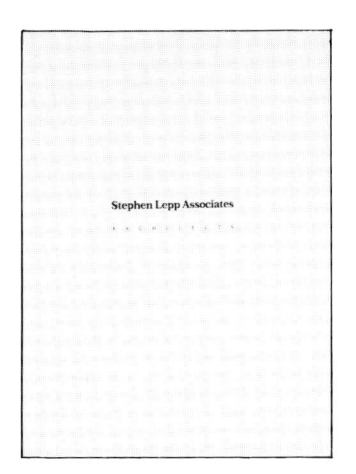

Stephan Lepp

A slick cream-colored cover with a subtle taupe grid and silver debossed type are used on the cover of this architectural firm's brochure. Different shades of this same taupe color are used throughout this piece for text, grids, graphic bars and as a tint in all of the black and white photographs. The use of this coloring and the repetition of the subtle cover grid help tie this brochure together graphically. A great deal of information, including quotes, the firm's philosophy, project descriptions and resumes of the firm's principals, is included in this piece, yet it still maintains an uncluttered appearance. Project photos as well as several renderings are found on every page and a list of selected commissions are supplied on the back of the brochure.

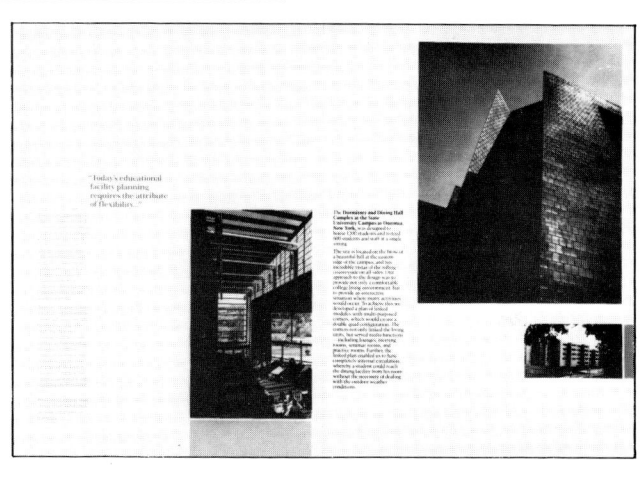

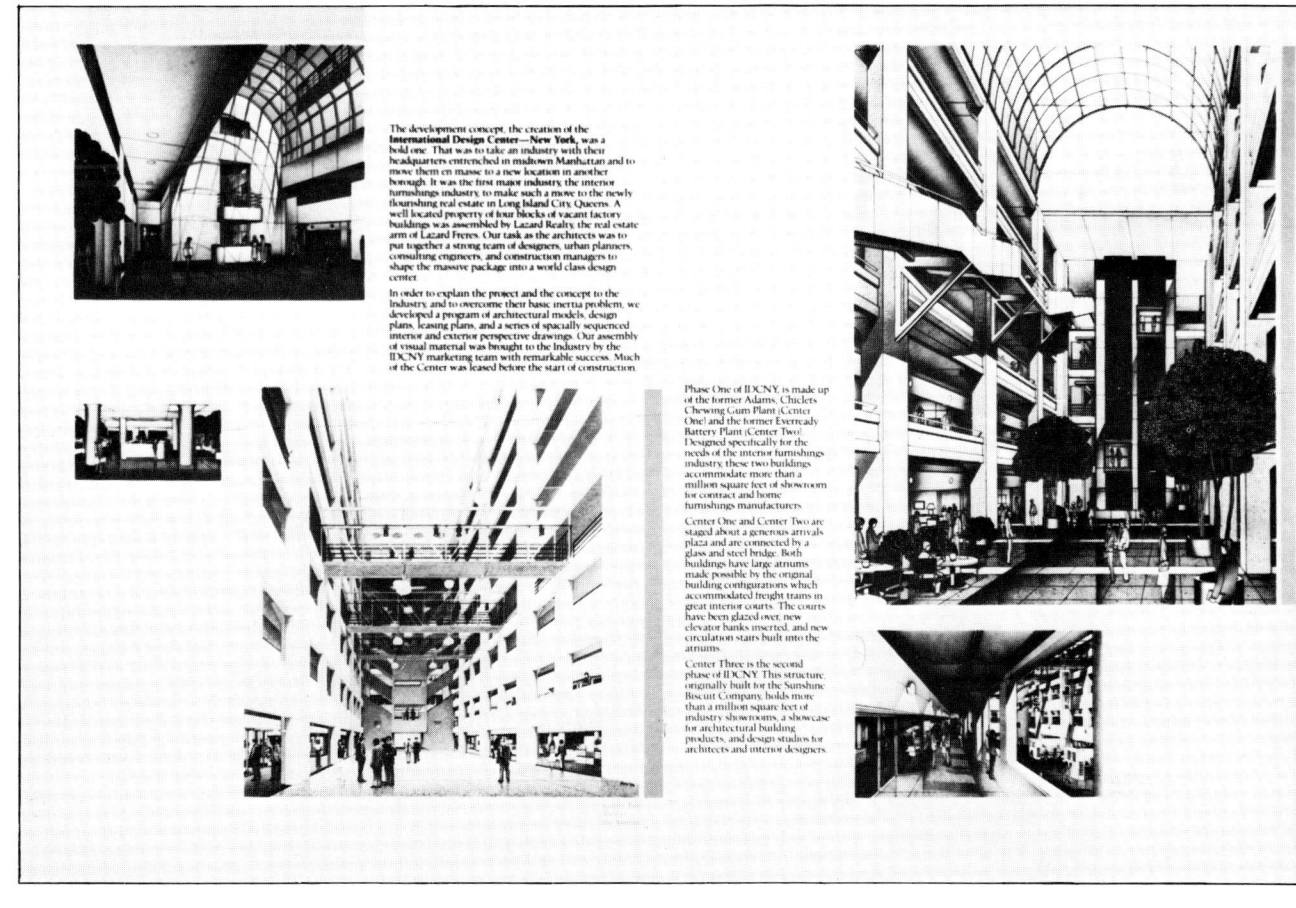

Company Brochures: Architects

TAC

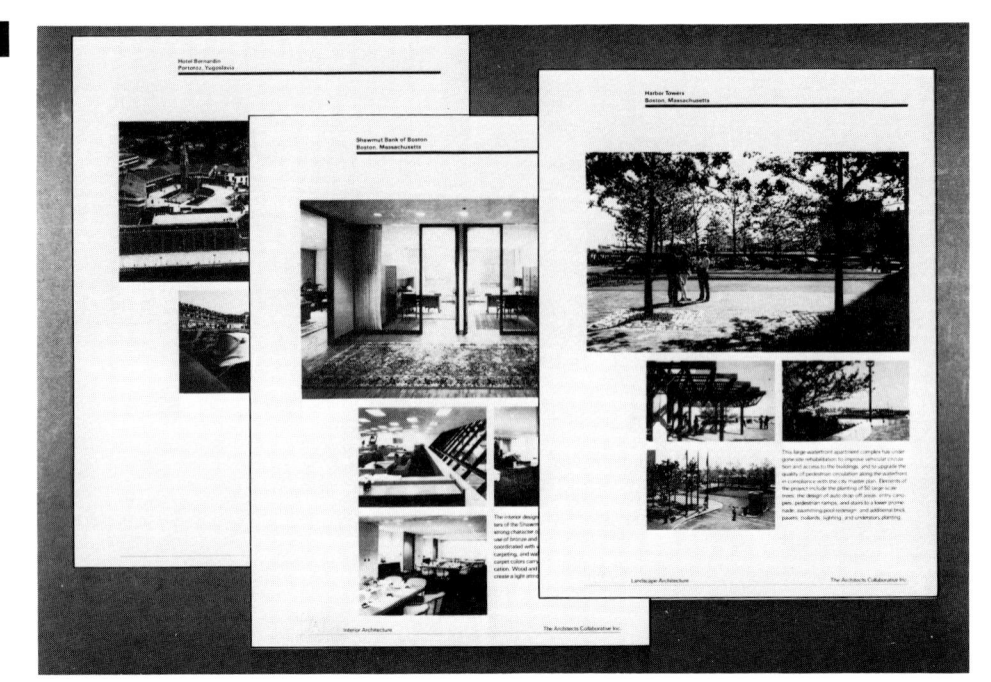

Project Pages

These one-sided pages show color photos of projects. The text is minimal and gives an idea about the project assignment and solution. Black rule lines are used.

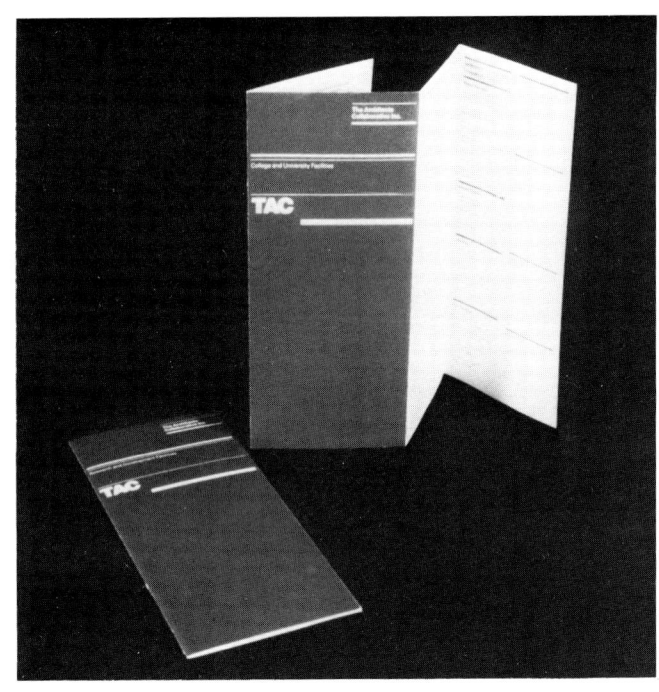

Specialty Folders

These pocket-size folders open out to reveal three-to-four-panel pieces. The covers, which come in a variety of colors, have white rule lines and show the name of the firm on the front. Some of these pieces have color photos of projects, and information about the firm.

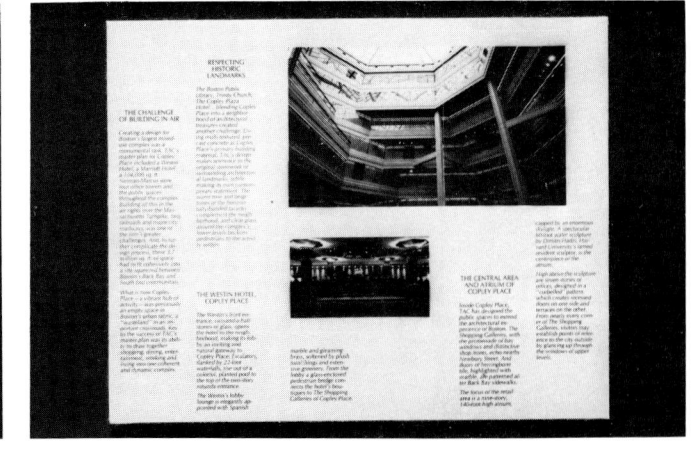

Hugh Stubbins

For its promotional material, this architectural firm has created a series of two-sided project pages. These single pages make use of full-color photographs which bleed from the right-side of the page. Black rules are used to further tie these pieces together visually. The reverse side of these pages show additional project photographs and text.

Project Pages

Project Folders

The firm also produces project pages that fold out. These folders use one large, full-color photograph on the cover and several detail photographs with text on the inside pages.

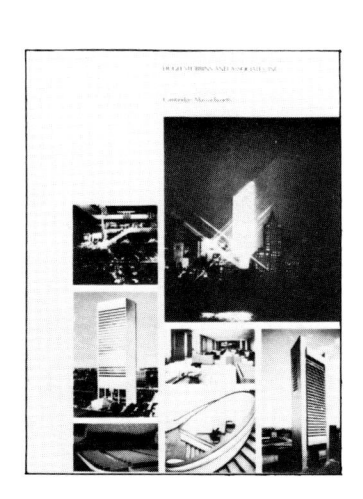

Company Brochure

This firm's company brochure keeps its design consistent with that of the project pages. The cover is a combination of projects the firm has been involved in, all shown in full-color. The firm's principals and employees have the spotlight on the interior pages. The text discusses the firms approach to its work.

Company Brochures: Architects 103

Anshen+Allen

This architectural, planning and interior design firm designed a brochure that gives the client a range of projects to view. The contemporary piece combines bold, brilliant color visuals with minimal text. The brochure is plum color where visuals do not appear. Unusual inanimate objects placed in pictures add a surreal aspect to the piece. The pages are labeled by different disciplines such as interiors, commercial, and educational. The brochure was designed in the format of project pages, wherein the back sides of the sheets were left unprinted on half the print run. This allows each page to be tailored to a specific proposal format.

PGAV

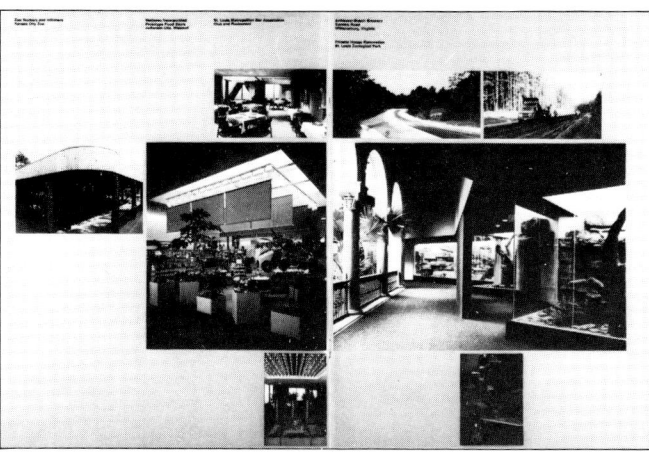

A glossy red cover with white type is a striking introduction to this firm's company brochure. The right-hand corner is clipped to reveal the company logo in black type against a light-brown paper. The inside cover page carries a client list. Each following page shows four-color photographs against a silver background.

Anderson DeBartolo Pan

The cover of this company brochure is glossy white stock with slate gray and red type. Gray rule lines are also used. Inside, four-color bleed photos appear on every page with minimal text about the firm and its goals. A wide blank margin across the top of the brochure is maintained throughout the piece. Some pages bleed onto the adjoining page. People are shown in many of the dramatic visuals, lending a human quality to the brochure. The back cover has "a final thought" about the firm's philosophy toward its work.

Company Brochures: Architects 105

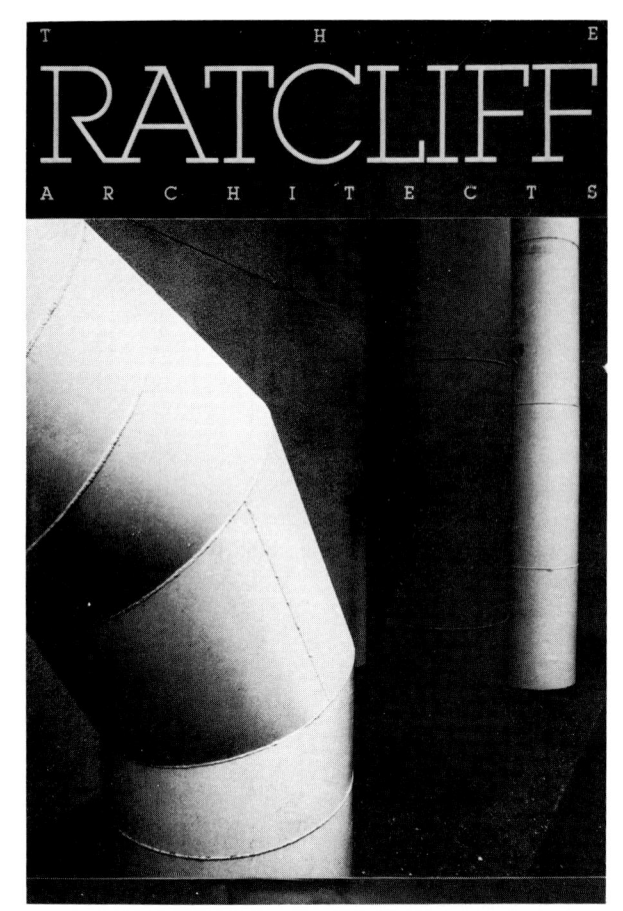

Ratcliff

This architectural firm has chosen company brochures that express a modern approach to design and architecture. Bold, bright colors (yellow, royal blue, orange-gold, and purple) are employed as rule lines against a gray background. This piece opens up to a gatefold, and the visuals are big and bold, showing details and unusual angles of projects.

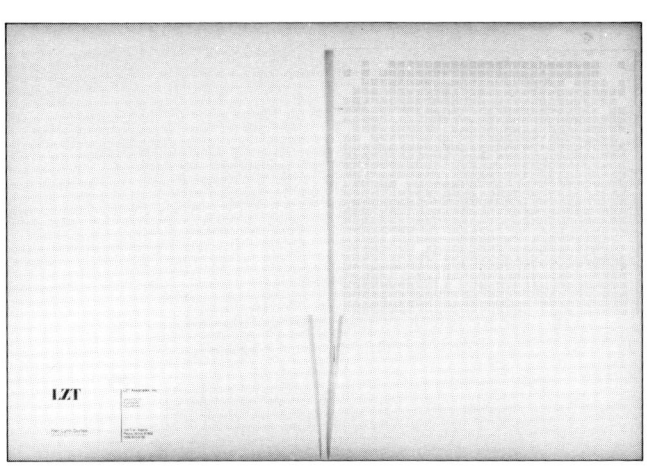

LZT

A four-color bound brochure was chosen by this architectural design firm. A subtle grid scheme was used to achieve a contemporary feeling. The beige-and-white grid, the name of the firm, and the company's logo appear on the front cover. (The company pocket folder also carries the same design.) Inside the brochure, gray type on beige, and beige type on black is used. Four-color photographs appear on a black field, and text is limited to the name of the project.

Company Brochures: Architects 107

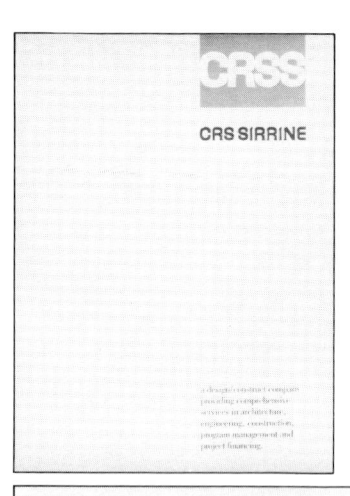

This design and engineering firm is a single corporation resulting from the combination of several entities which were acquired by the parent company. The logo typeface reflects this merger clearly. On the front of the company brochure, an orange and white logo against a white field highlights the logo type. Beneath this identity is the name of the firm as it currently stands. The clean-looking cover has only a brief description of the services offered by the firm. Inside, four-color and black-and-white photographs accompanied by brief text (discussing the firm and its projects and services) appear. Throughout the piece, blocks of type are highlighted in black and white to emphasize the point. People are shown at work frequently throughout the piece, adding a human quality to this technically oriented piece.

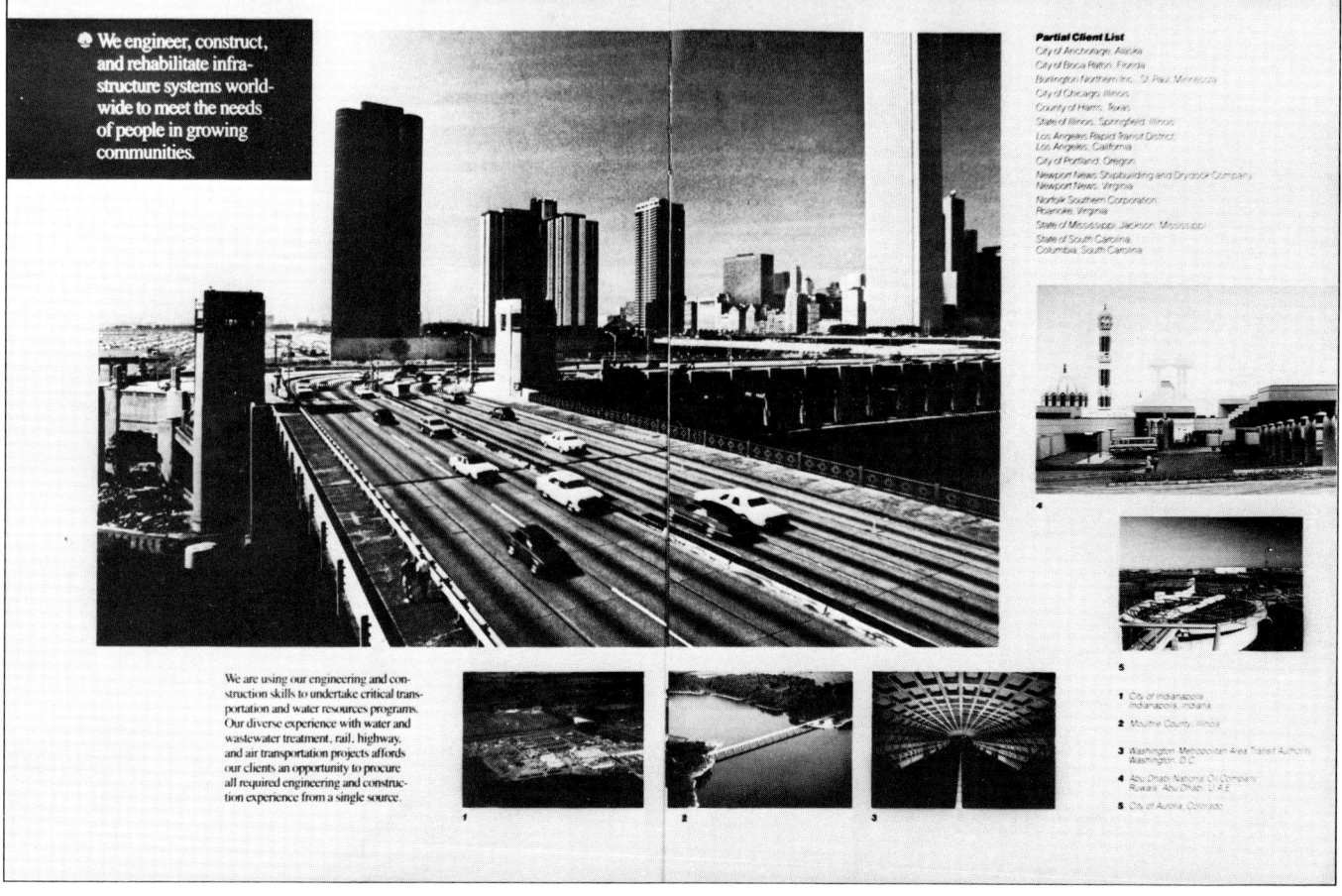

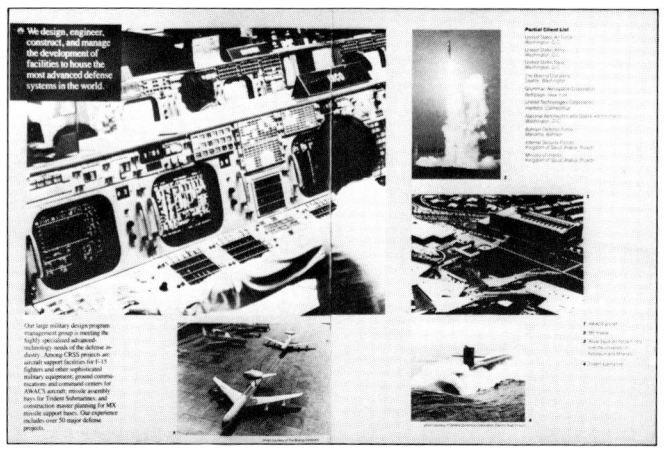

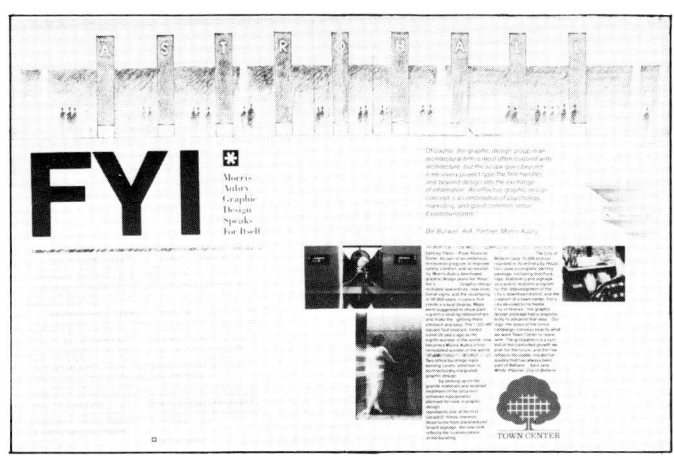

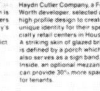

Morris * Aubry

The company brochure for this architectural firm uses a glossy white stock. A color photograph and the name of the firm appear on the cover. The format of this brochure is bold throughout, without using a specific grid structure or repetitive page design. The type is clean and simple, yet is intentionally different from page to page. Inside, the reader is introduced to the firm by a letter from one of the partners. Text, color photos, and renderings appear throughout the piece. Quotes from company executives appear with the photographs.

Company Brochures: Architects

The Kober Group

This architectural firm uses a three-fold format for its company brochure. Using a unique six-column square grid system, it takes full advantage of all of the space on the page and still maintains an uncluttered appearance. A matt gray finish is s used throughout this piece and all text is dropped out in white. Graphic bars are also used and they appear in shades of red and orange. The full-color photographs used are all keyed to lists of project locations and titles. The text is kept to a minimum and discusses the firm's philosophy and experience gained in its 50 years of business. A listing of corporate offices is supplied on the back panel.

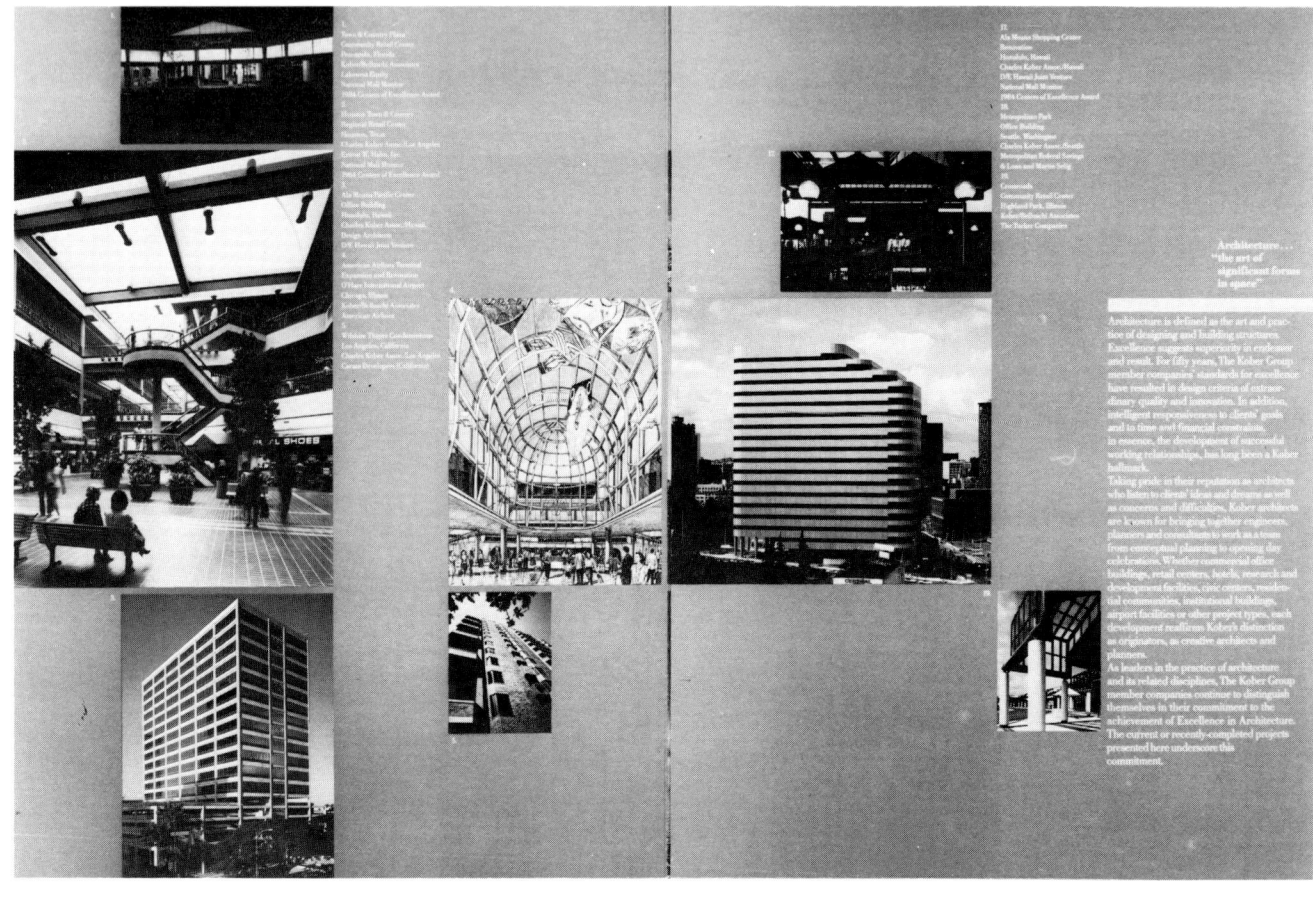

Sugimura

This architectural firm uses a four-page folder style brochure. The format is kept very simple and the brochure is laid out on a diagonal grid that is four squares across and five squares deep. The innovative way this grid is used gives this piece its sharp, well thought-out appearance. All photographs and text fall within these tilted squares, including the front cover logo and the back cover office photograph and address. The text discusses the firm's method of operation and company philosophy. Project titles and locations are also given.

Company Brochures: Architects

Flack + Kurtz

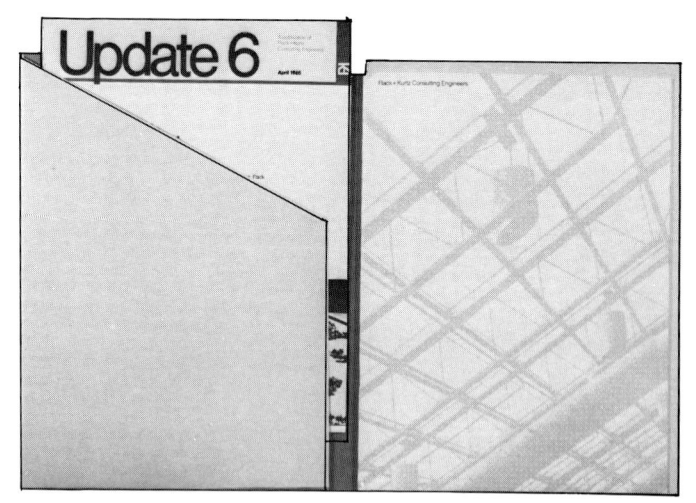

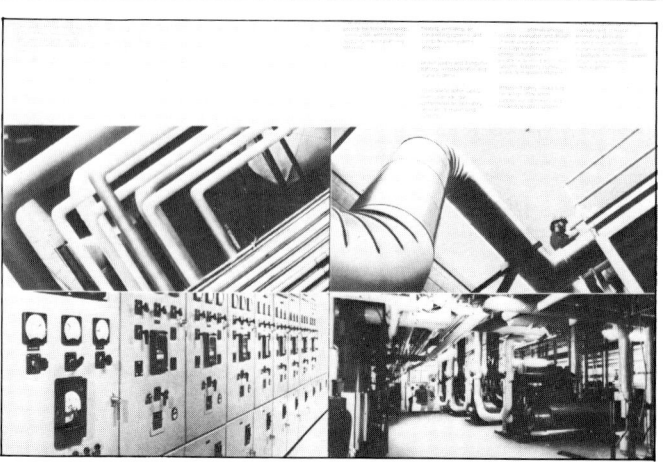

This consulting engineering firm uses a folder with one pocket and one flap to hold its brochure and newsletter. The graphic used on the cover of this folder is a coarse line screened photo reproduction of one of the company's projects.
To make the image even more subtle, a gray on white color scheme is used as opposed to a black and white one. The company's name is found in the upper left hand corner in small red type which is easily readable but does not interfere with the cover graphic. The same gray is used again for the inside flaps while the remainder of the inside matches the red used for the company's logo type.

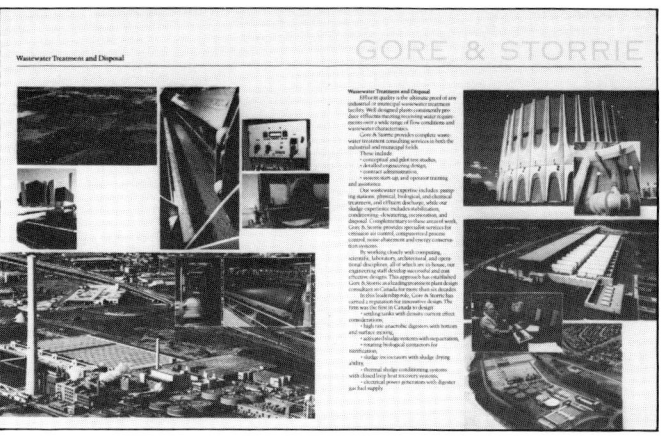

Gore & Storrie

This consulting firm specializes in environmental engineering. The full-page cover photo shows clear bubbling water, representative of the firm's field. Inside, color photos are seen on every page. Color insert photos bordered in white, which show close-up views of the firm's employees and projects, are an important part of the design.

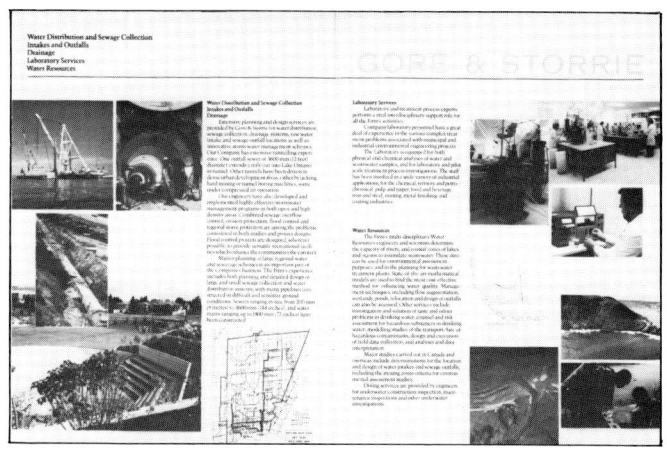

Elson T. Killam

This engineering firm uses inserts to give either an overall view of a project or to pinpoint a detail. Bordered in white, these photographs are inserted into full-page bleed photographs. Though these insert photographs get attention, they do not draw interest away from the larger photograph.

Company Brochures Engineers 113

Culp • Wesner • Culp

With this firm's ten-year face-lift came a new company brochure. The goal of the brochure was to convey that the firm is a full-service consulting engineering company. The targeted auddience was primarily public and private officials rather than practicing engineers. Color photographs with people as subjects were relied on, rather than lengthy text about treatment plants. The brochure's cover is black mat on heavy stock. A color photograph in orange and yellow and the company's logo and name also appear. A red rule line across the top of the page and beneath the type is repeated throughout the entire piece. The piece folds out into a three-page spread.

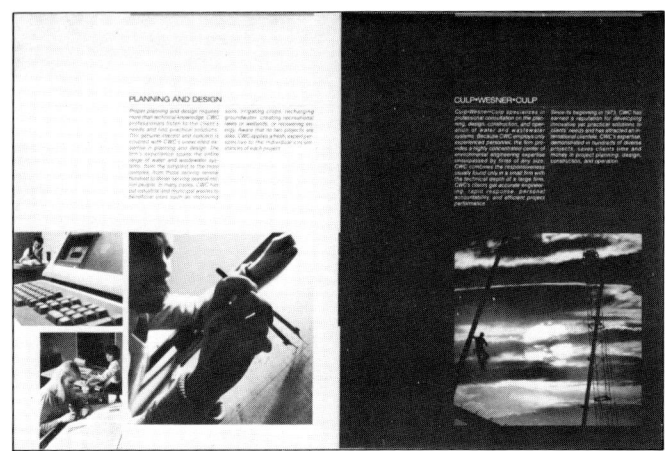

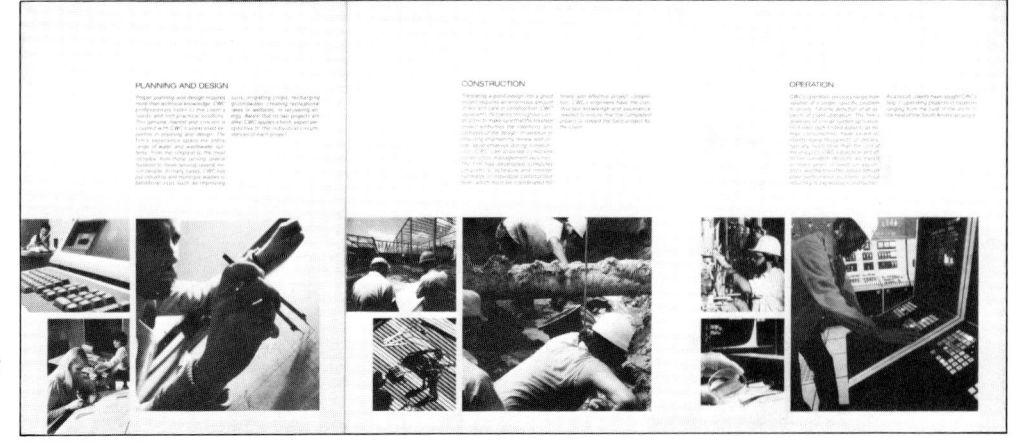

Each page is composed of four-color photographs which appear across the bottom of the page. Above the visual is brief text about planning, design, construction, and operation.

114

Woodward-Clyde

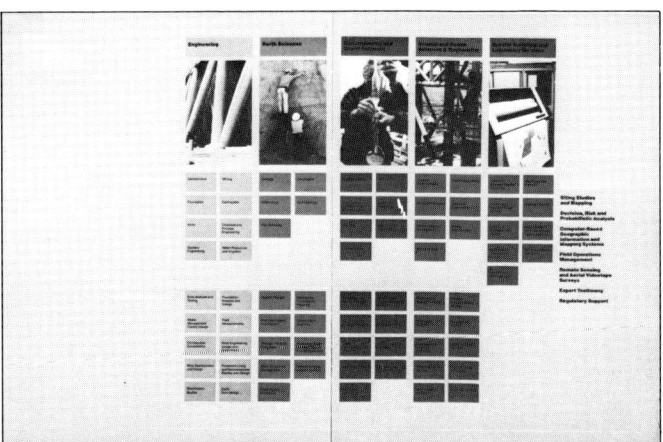

Concern for the environment is emphasized in this consulting engineering company's brochure. The cover of this piece is a dull varnished gray color with a vertical die cut. The narrow die cut reveals a portion of the photograph featured on the first page. The colors exposed through the cutout are vibrant shades of blue, green and orange. Black type was chosen and creates a straightforward look. Inside, four-color photographs appear with text, which describes the firm and its approach to civil engineering. Pages are organized according to specific categories and color photographs appear in a grid-like formation. People, buildings and construction sites are among the visuals used. At the back of the brochure, a client list is provided as well as a pocket where additional materials can be stored.

Company Brochures: Consultants 115

Emcon

This consulting firm manages waste and environmental control. It designed a four-color brochure. The outside photo of drops of water on blades of grass was chosen to depict a pristine environment. When open, the brochure reveals four photographs which cover each of the firm's four areas of expertise

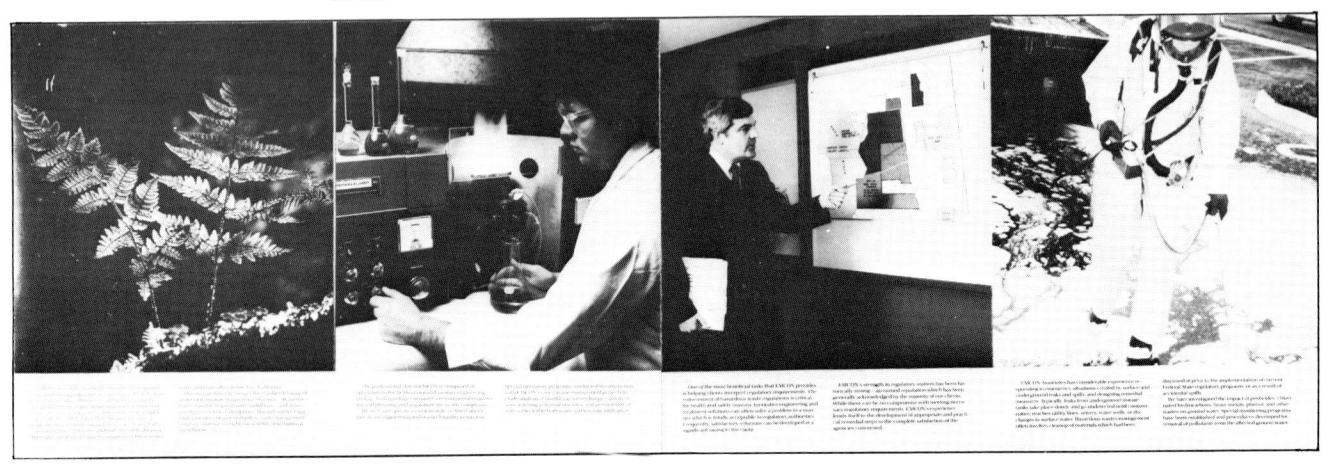

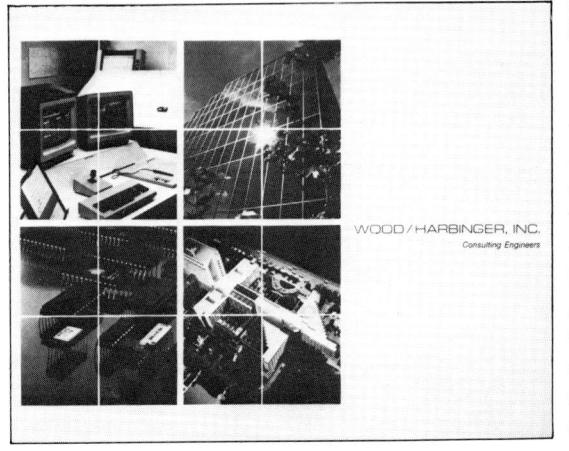

Wood/Harbinger

This consulting engineering firm uses a horizontal format for its company brochure. The brochure's cover shows four photos of company facilities and completed projects. An embossed white-on-white logo as well as the company's name and discipline, appear here also. Inside, the brochure is broken down by the firm's divisions. Each division section features full-color photographs of finished projects or models. The photographs use the same grid pattern as the cover. Charcoal gray rules are used throughout the piece. A client list is provided in the front of the brochure.

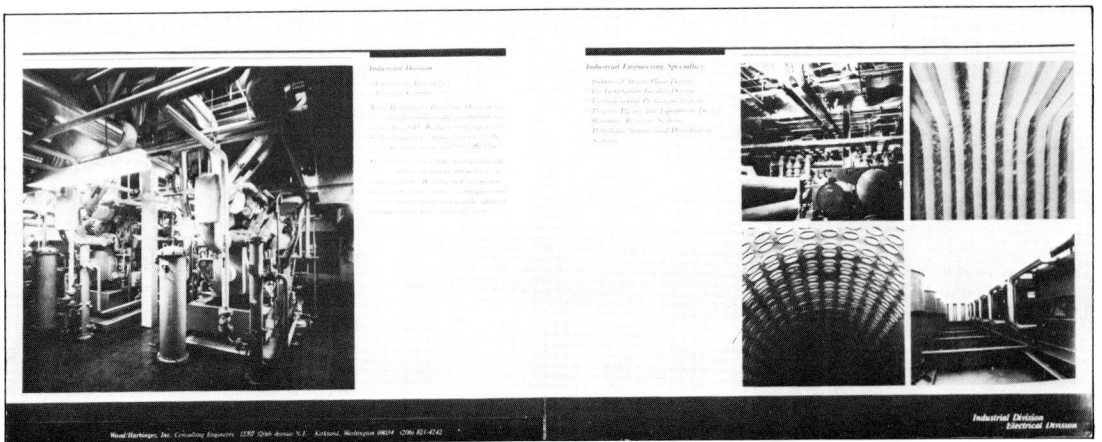

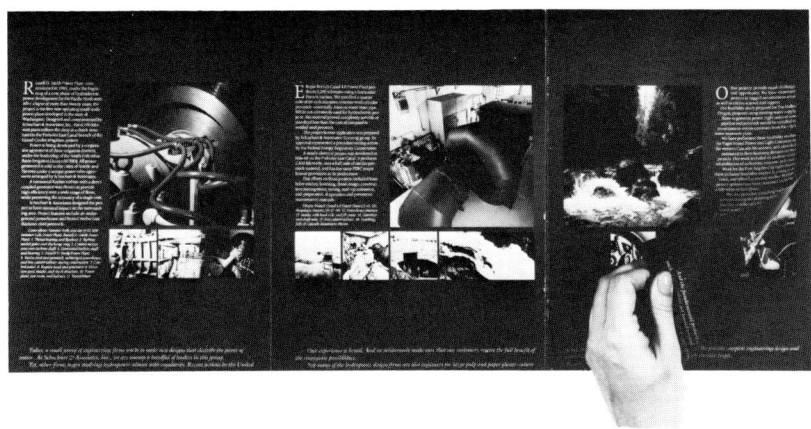

Schuchart

This consulting engineering firm has created a brochure which covers its main specialty, water power. A dramatic color photograph of water in motion is used on the cover. The photograph and the bold words "Water Power," along with the name of the firm, are set in a blue background. The inside front cover which continues the dramatic photograph from the front cover and gives an interesting bit of information, leading you in to read the rest of the brochure. The inside cover also folds out to reveal color and black-and-white photographs of completed projects. A column of text is used on the left side of the piece and discusses the firm's projects. Peach-colored type is shown along the bottom of each page, and this text discusses the firm's track record.

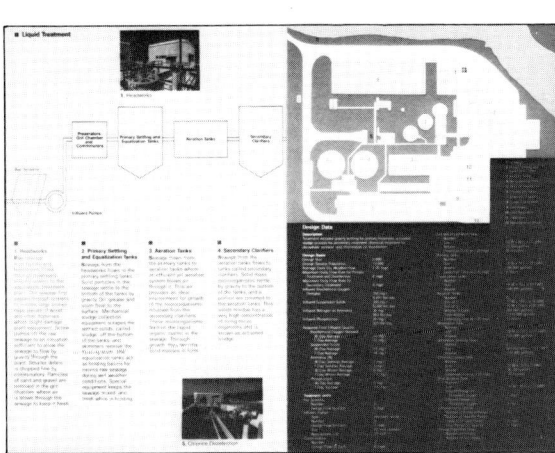

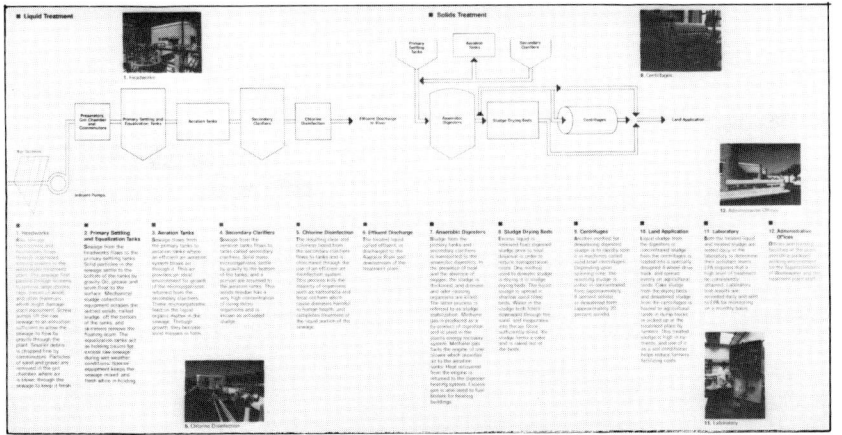

Brundage Baker Stauffer

This firm of consulting engineers uses a graphically-matched brochure system. The specialty piece, featuring a wastewater treatment plant, uses a three-panel foldout format. This format works well with the project flow chart supplied inside. A numbered layout of the plant is shown which corresponds with the numbered descriptions given across the paneled inside. Full-color photos of the completed project are scattered throughout this piece. These numbered photos also tie in with the descriptions and plant layout. This system of numbering gives you a clear picture of what was involved with this project.

Company Brochures: Engineers 117

NAB

The cover of this dramatic brochure is a strong red metallic stock with embossed logo and silver lettering. Photographs of the firm's corporate offices, accompanied by brief text about the firm's history and philosophy, create a personal tone. Full-page four-color spreads illustrate the company's vast industrial scope.

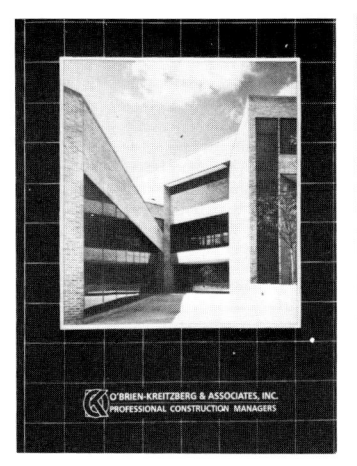

O'Brien Kreitzberg

This firm specializes in construction management. Its company brochure aims at familiarizing the prospective client or continuing client with the firm's experts and the professional services offered. Throughout this twelve-page brochure, the grid design shown on the cover is maintained in black and white. The brochure begins with a letter from the firm's CEO and president. A list of company services is supplied, as is a roster of its top executives, with their photos and resumes. The centerfold is the only area where four colors are used.

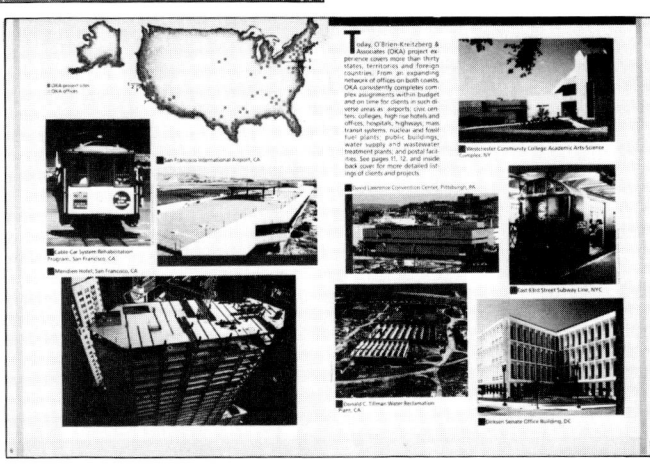

Brae

This design and building team uses a deep-blue front and back cover with white lettering, company logo, and rule line. The white rule line and logo are continued throughout the brochure. Four-color photographs of people who work for the firm, and of site locations, are featured on each page. The text specifies what services the firm offers and its philosophy and approach to construction. The back cover carries the addresses of the firm's two main headquarters.

Company Brochures: Construction 119

Tishman

Two main brochures represent this construction company. Though both pieces use the same format, they remain graphically individual. The piece folds open to a gatefold, revealing a high contrast graphic of a building. Then, when opened again, it reveals color photographs set against a black field. Blue type is also set into this black field.

The second brochure has the same gatefold format, but the tone is softer using blues and grays as the main colors. The white rule lines used on the cover are continued in black inside the brochure. Buildings are shown against a white field. Minimal text is used.

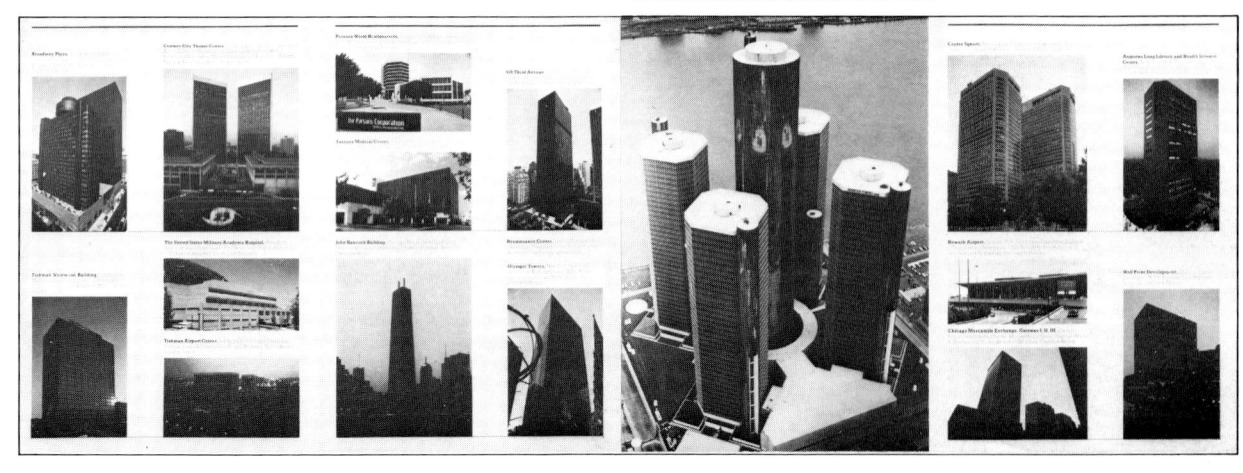

Partners

This construction firm specializes only in interiors. It created its company brochure to convey in one minute or less a positive image, one of creativity and high-quality professionalism. The front cover is teal blue with a pale gray logo and graphic. The back cover also carries the firm's logo and its name and address.

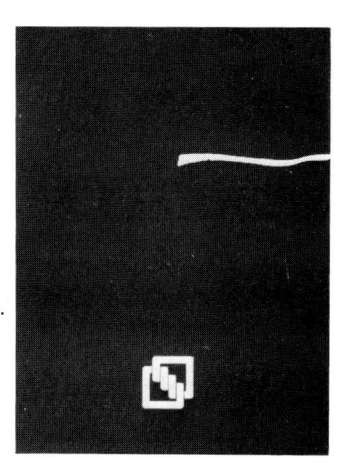

Company Brochures: Construction 121

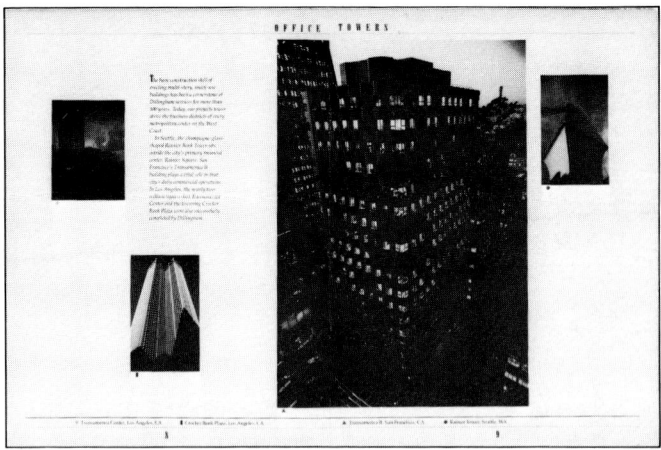

Dillingham

Though the two brochures shown here are from the same construction firm, the design and overall impression they give is quite different.

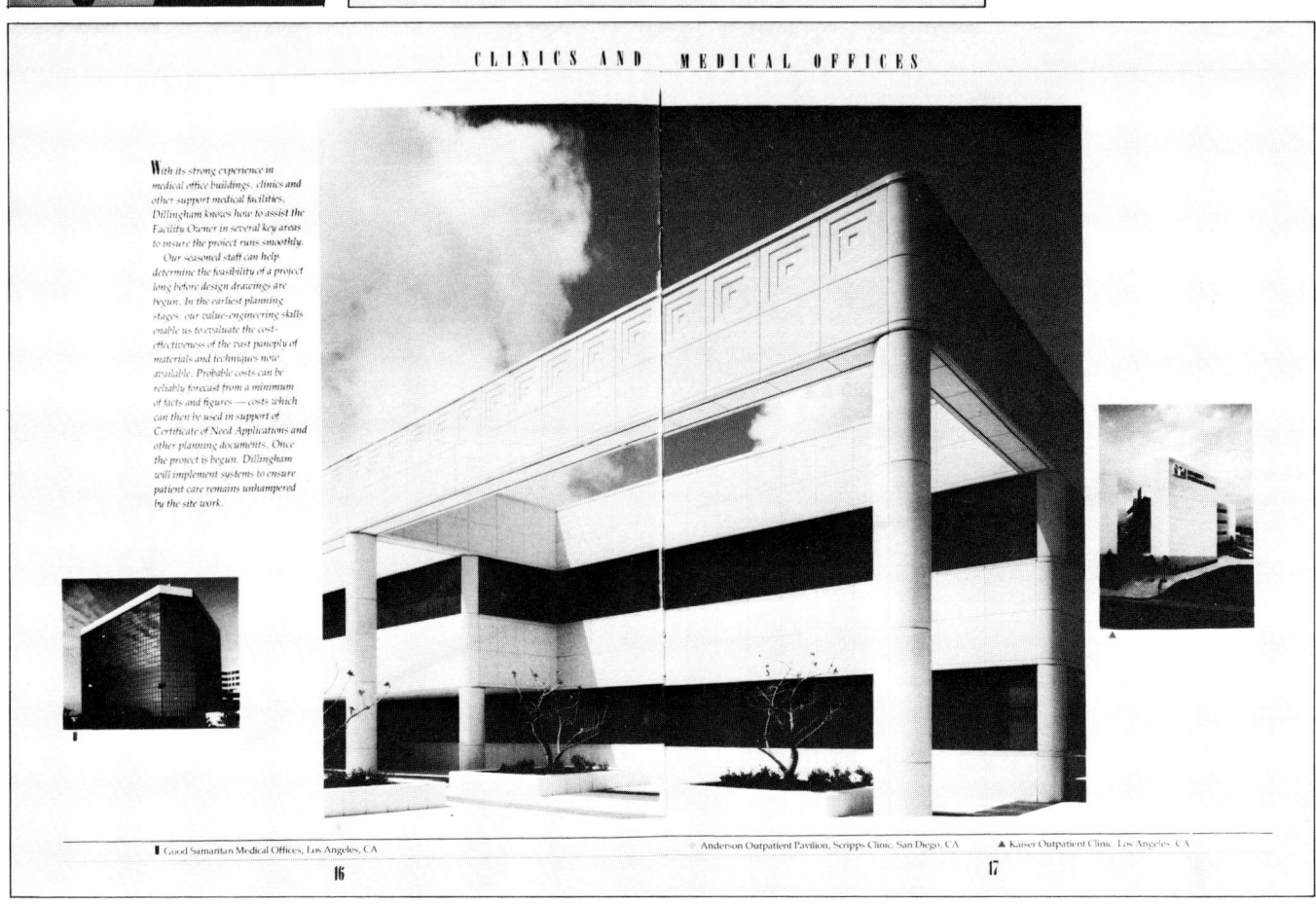

The architecture brochure uses a silhouette of a construction worker for its cover, which creates a strong graphic image. The four-color photographs used throughout this piece are striking project views. Large photographs, (which bleed across the gutter) dominate the pages. Smaller project photos are used as well and all photographs are keyed to the bottom of the page with the aid of geometric shapes.

This brochure identifies the firm's services. The cover of this brochure also shows a construction worker precariously perched on a project in progress. Large full-page bleed photographs which relate to the service discussed are shown on the left-hand side of the spread. The right-hand side contains a large headline identifying the service, a smaller photograph, and a column of text describing each service.

Company Brochures: Construction

GHK

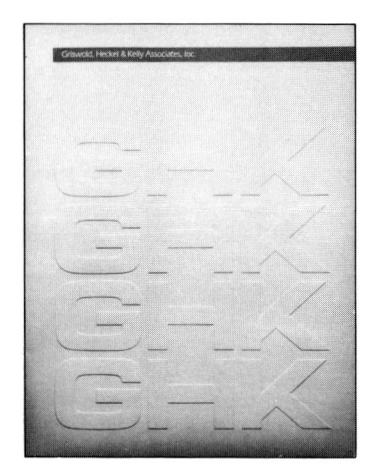

This firm specializes in interior design, planning, and architecture. It's marketing materials have been designed as an integrated brochure system. Three separate designs are used for the covers of the pieces. These graphics are based on a design motif of the square within the square. The inside of each piece features one large four-color photograph with brief text. The main company brochure uses the same format as the other specialty brochures, but includes more pages and text. The text discusses the firm's capabilities and philosophy towards it's work. There is a flowchart which diagrams the firm's design management process, detailing how the firm handles a client's project. All material can be selected to fit each individual client.

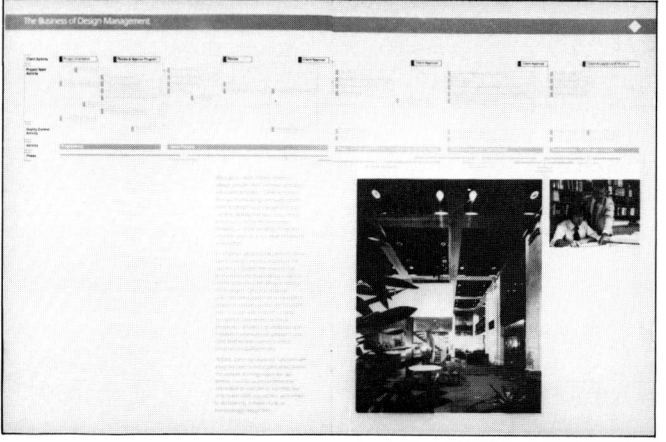

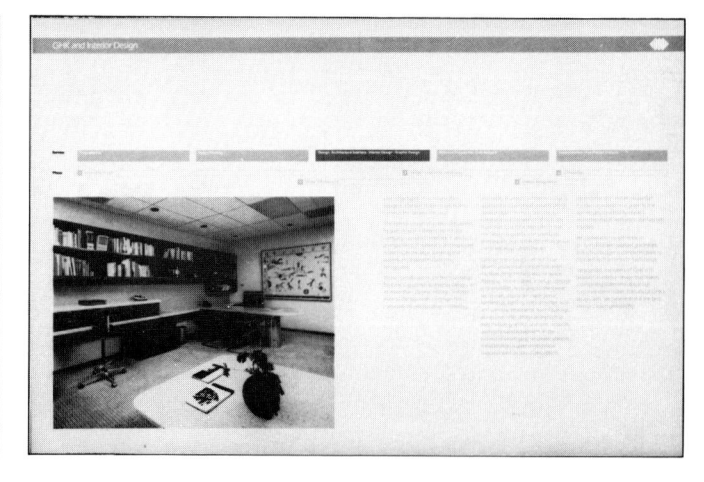

Trisha Wilson

This interior design firm has created a brochure system for its corporate marketing purposes. A red lacquered portfolio-style pocket folder held closed with a red string holds all of the material inside. The name of the firm appears in black type at the bottom of the folder. The folder opens to reveal a black lacquered interior. On the left there is a pocket with eight insertable pamphlets, each describing a different facet of the firm. The center of the folder has a larger pocket with 8 x 10 inch glossy photographs and reprints of published articles. A descriptive adhesive label is applied to the back of each photograph, giving further information. The flexibility of this format enables the firm to personalize each portfolio.

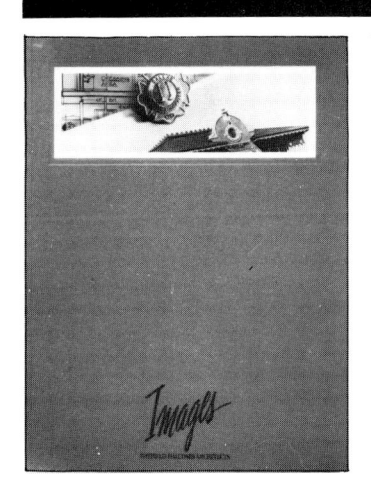

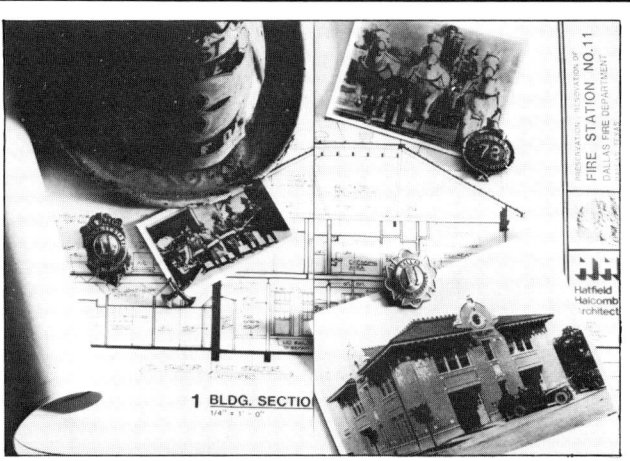

Hatfield Halcomb

The cover of this company's brochure is a steel gray heavy stock. The embossed area near the top of the piece is used to place a close cropped photograph. Since this photograph is interchangeable, the brochure can be personalized for different target markets.
Inside this brochure, black mat pages with gray type and the firm's logo appear. These pages are used to separate different sections of the portfolio. Collages are used near the front of the brochure to highlight each target market, and, like the cover photo, are interchangeable. The other pages feature large vibrant color photos of projects. The text is kept to a minimum in favor of these visuals. The pages that show projects can also be used alone as project pages.

Company Brochures: Interior Design 125

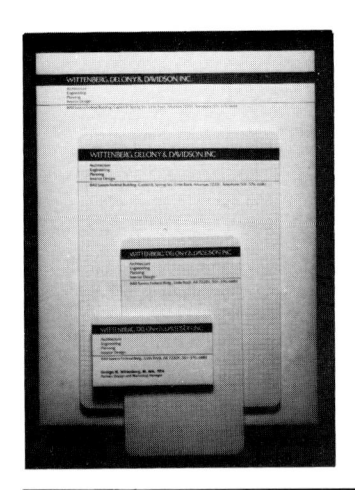

Wittenberg, Delony & Davidson

This engineering firm devised a brochure system for its marketing purposes. All brochures and project pages are contained in a two-tone beige folder. All marketing materials for this firm are kept graphically simple with the same beige-on-beige color scheme. Rule lines in various colors are used to differentiate these pieces and all type used is white, and simply states the name of the firm and the brochure specialty area.

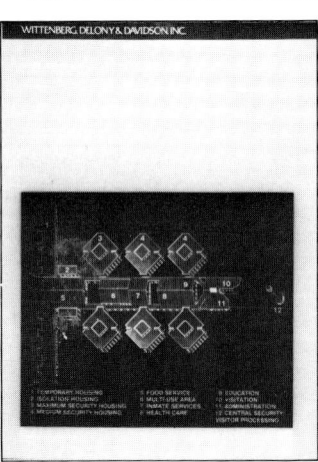

Project Pages

The one-sided pages have a red rule line. A thin black line separates the top portion of the page, where there is text, from the bottom half where there are four-color visuals of the project. The visuals include floor plans, renderings and photographs. Some pages use yellow or blue rule lines.

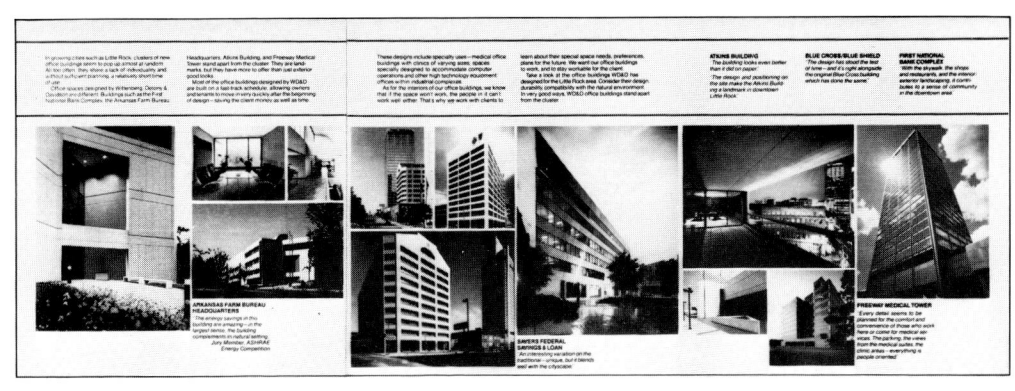

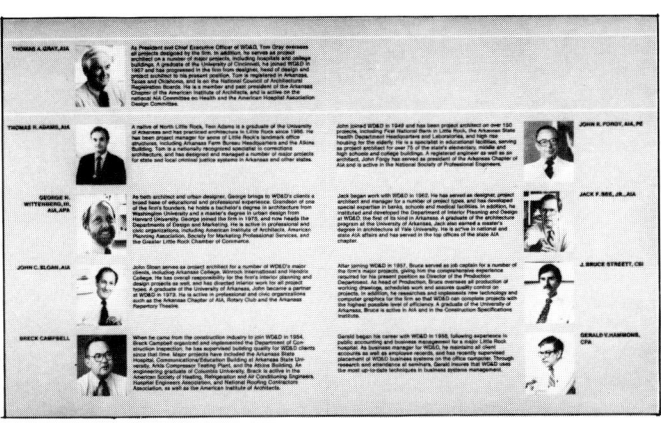

Specialty Brochure

The three-page fold-out pieces are shorter than the main company brochure. The brochures use the same two-tone beige combination and color photos on the inside. The layout is the same as the rest of the materials, with the text at the top and the visuals at the bottom of the page. Quotes are an important part of this piece.

Company Brochure

The company brochure is taller than the specialty brochures. The outside features double red rule lines. These lines are continued on the inside and the back of the piece. The red rules on the inside separate the text at the top of the page from the visuals at the bottom of the page. Black-and-white photos of company executives are used at the top of the page along with text.

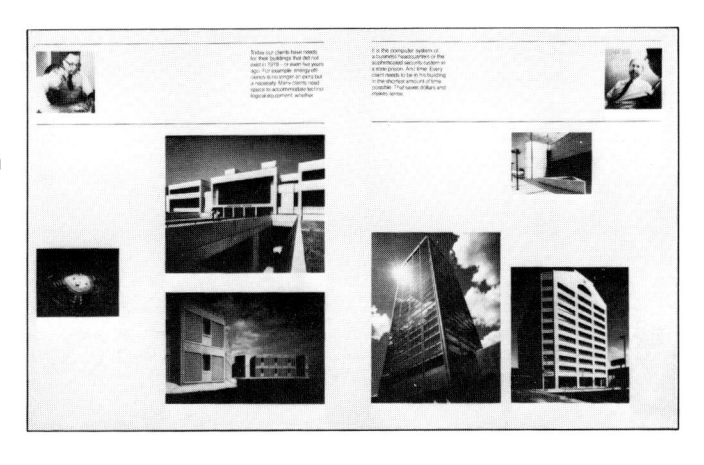

Company Brochures: Architects 127

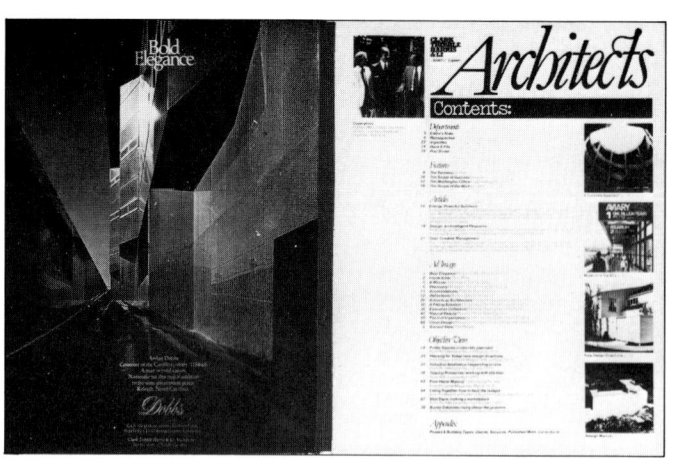

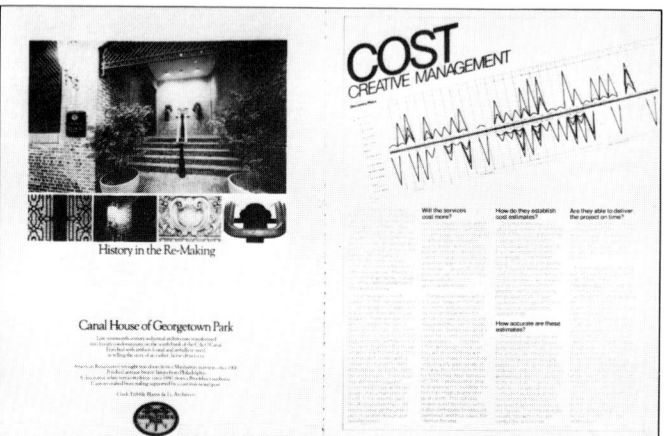

Clark Tribble Harris & Li

This architectural and engineering firm has produced a series of brochures for its marketing purposes.
The first piece looks very much like a magazine. It is printed on magazine-type stock, and uses a table of contents to list articles. The typefaces and layouts vary from page to page. The advertisements, which appear to be standard magazine ads, are in actuality all ads for the company itself. The articles are also about the firm. They discuss recent projects and the company's philosophy.

The second brochure in this "continuing series," as the company calls it, resembles a newsletter. The two-color piece is divided into several sections. These sections highlight the firm's area of expertise. Among the sections are pages that have been laid out like a collage, with several black and white renderings of buildings.

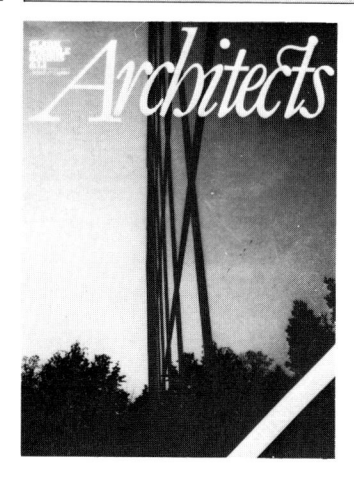

The third brochure in this series is done primarily in a mat black finish. The cover is mat black with black gloss type. The firm's name is shown in white. Dramatic full-color, full-page photographs of completed projects are used throughout this piece. Text is kept to a minimum, in favor of these photos. A unique project/client list that includes renderings is provided at the back of this brochure.

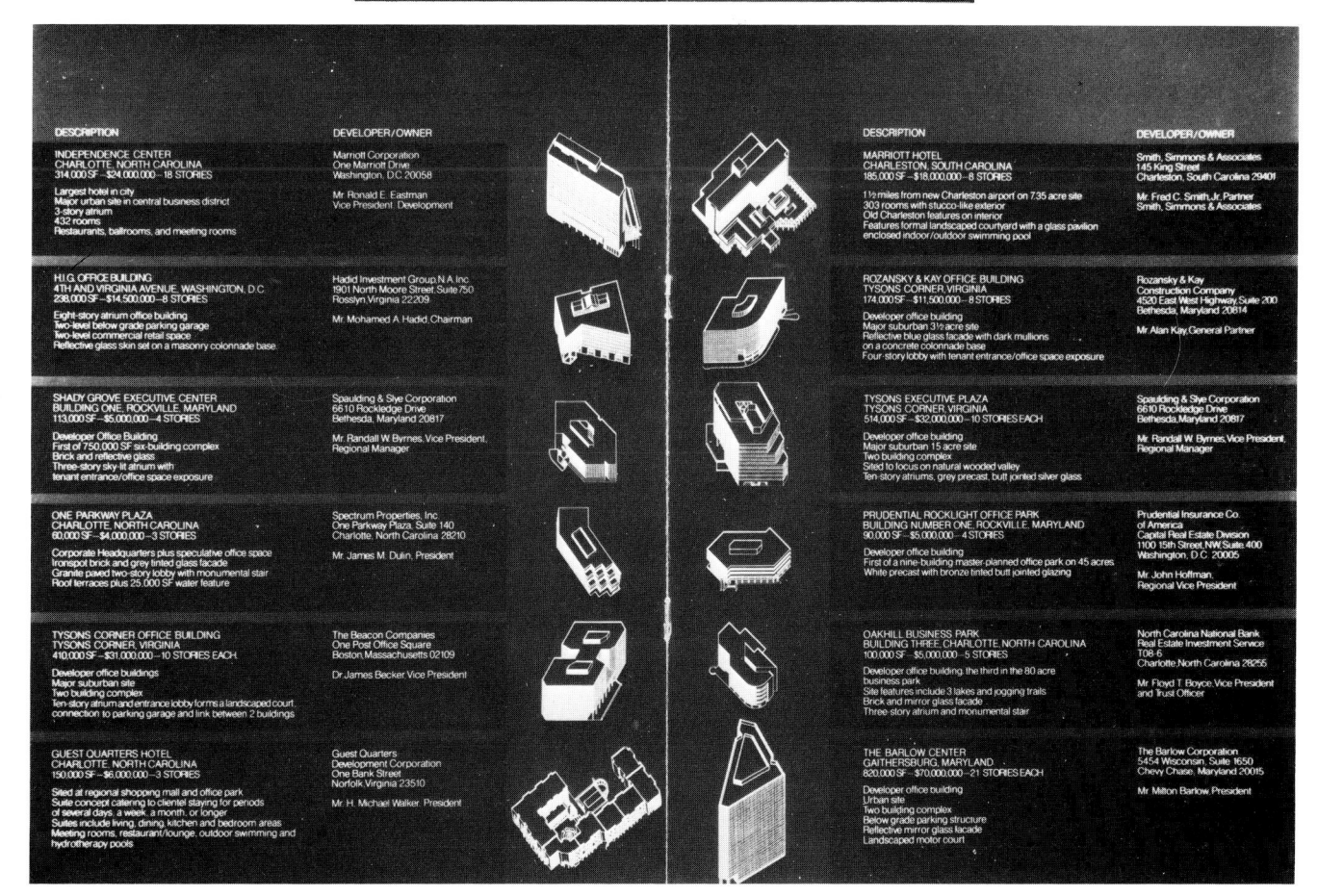

Company Brochures: Architects 129

The Hillier Group

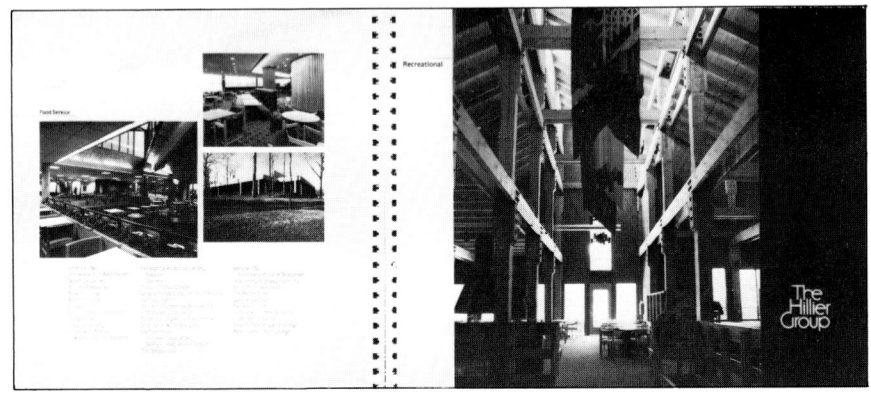

This firm, which specializes in architecture, planning, and interior design, has adopted a brochure system consisting of a company brochure and additional specialty pieces. The spiral-bound company brochure consists of forty-six pages designed in a horizontal format. The front and back cover are sheets of smoke-colored acetate. The interior is separated into the company's diverse professional disciplines. Color photographs and floor plans add to the visual impact. The text is limited to brief explanations about the firm and the projects. The three-page foldout and individually colored specialty pieces all carry the company logo (the firm's name) in a lined square, and a cover photograph, also set into a ruled square. Inside, color photographs of finished projects and brief captions appear. On the back of each of these pieces are personnel data, awards, and new commissions.

WTW

Single sheet pages are used by this architectural firm. Full-color photographs of projects (overall and detail) are used. The name and location of the project is seen at the top of these pages. Text is used to outline the details of each project and a fact list is shown.

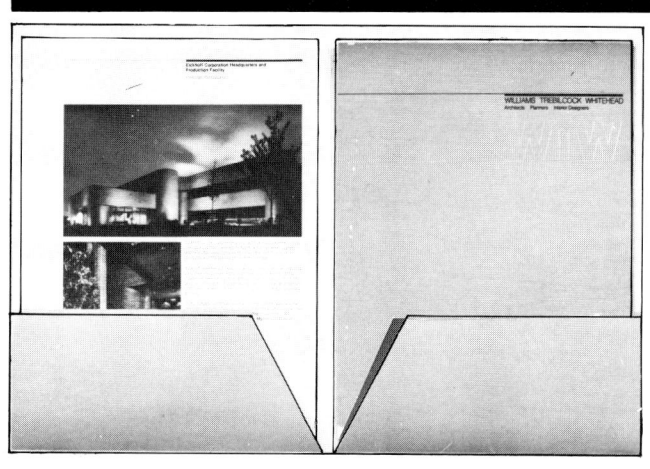

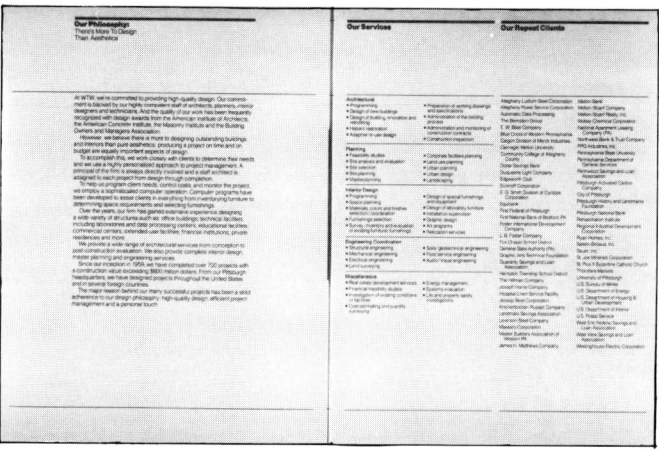

Folder/Project Pages

A glossy tan-and-brown cover printed with the firm's name on it is used to hold project pages and a gate-fold brochure. The call letters are shown as a blind emboss on the cover. The inside covers have pockets that hold all of this material This gate-fold brochure has a matching cover with the folder. When opened, three sections of type are shown, each covering a different subject. The firm's philosophy is on the lefthand side and a service list and client list on the righthand side. The piece then folds out fully to reveal full-color photographs of selected projects. Each photograph is numbered and keyed to its identity at the bottom of the four panels.

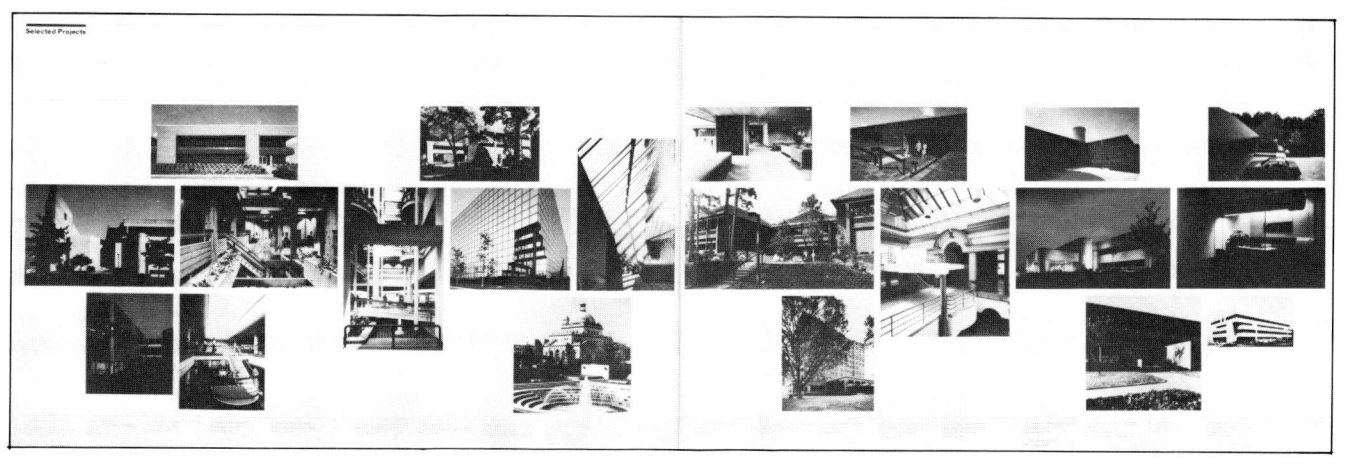

Company Brochures: Architects 131

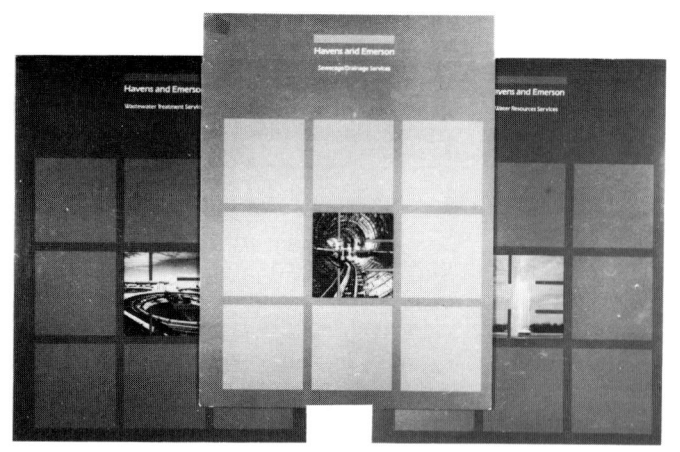

Havens and Emerson

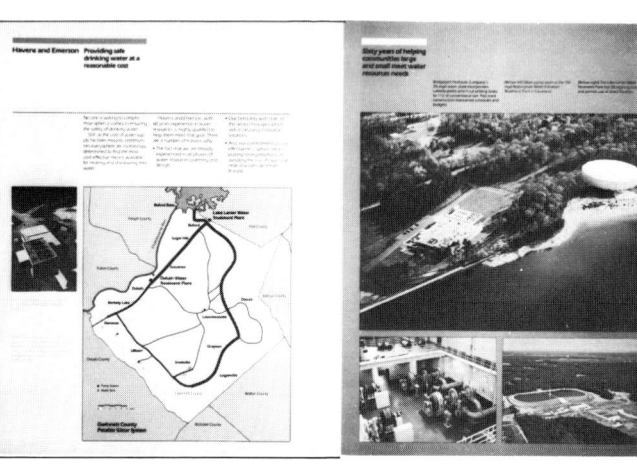

Specialty Brochures

These three-page foldout brochures are printed on heavy stock. The cover carries the same company logo as the corporate brochure, against a grid background. Each specialty brochure has the same grid design, but uses a different color scheme. The text highlights the company's history and expertise, as well as its philosophy about its work.

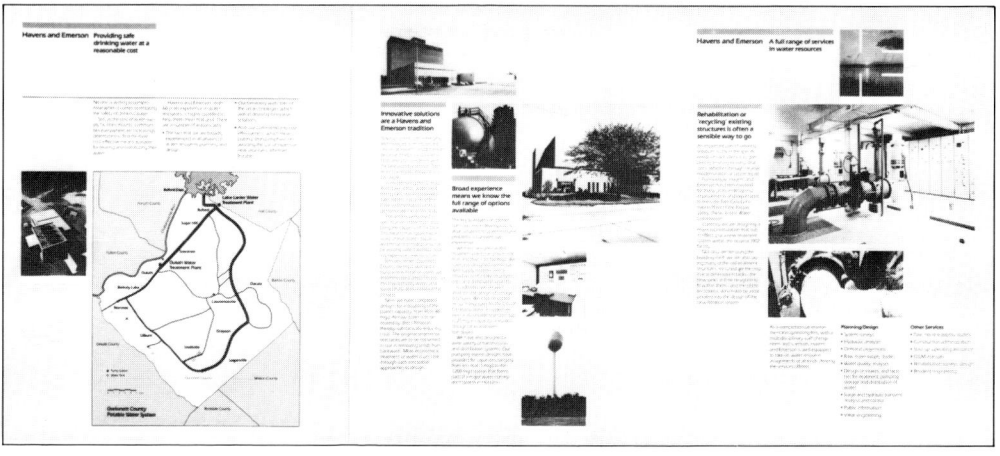

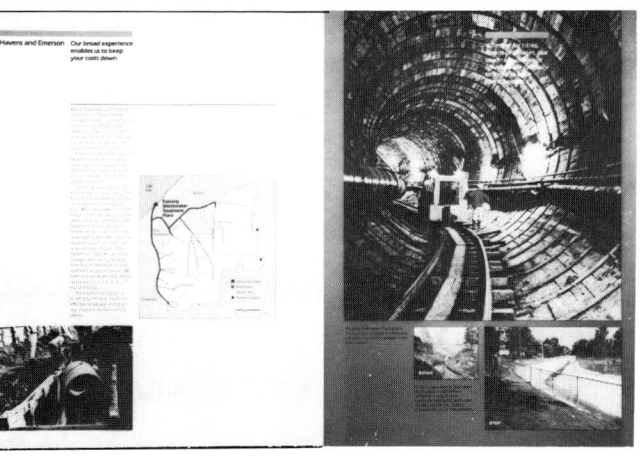

Company Brochure

The main brochure has a glossy white cover with a gray rule line, color logo, and photograph (which is part of the logo). Introductory text about the firm is printed on an acetate overlay. Full-page color bleed photographs are used as well as smaller cropped photographs to show details of projects and highlight aspects of the firm.

Burgess & Niple

This multidisciplinary engineering and architectural firm chose to use two company brochures. The main brochure has a moss-green cover with diagonal silver stripes. Inside, a frosted vellum sheet introduces the reader to the brochure. On this sheet, the silver diagonal stripes are repeated, while the firm's services appear within the stripes.

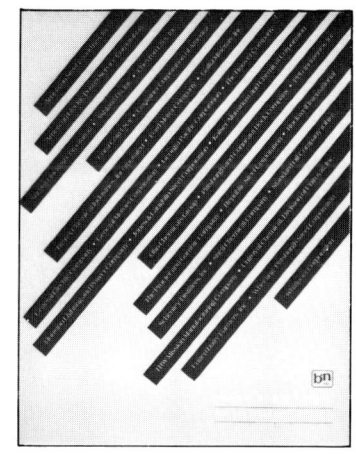

Services Brochure

The services brochure also employs the diagonal design; the services appear in the green stripes. The firm's name appears in white, while a green logo and green rule line pick up the green stripes. The piece folds out into a three-panel spread. A thick green rule appears at the top of most of the black-and-white photos. One page gives a brief biographic resume of some of the executives with the firm. The back cover is a repeat of the main brochure, except that the colors have been changed.

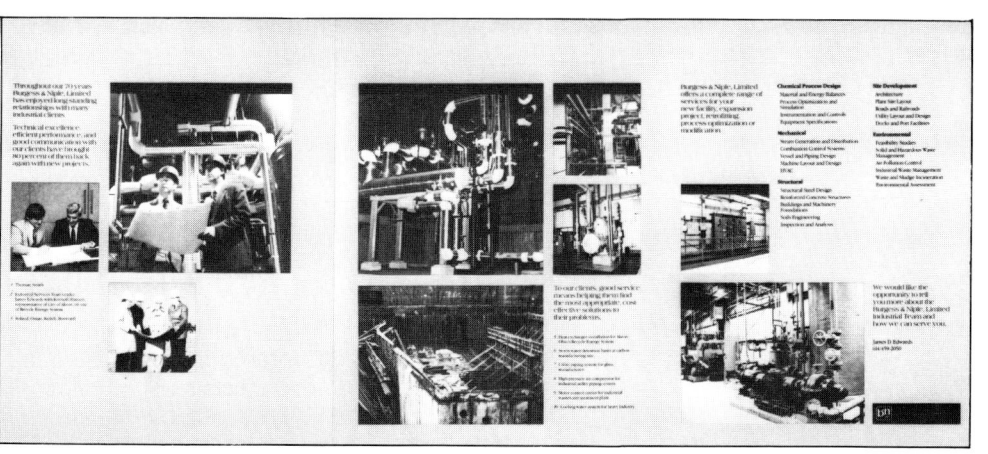

Company Brochure: Engineers 133

Elson T. Killam

The brochure system for this engineering firm was designed so as not to become outdated. Therefore, pictures of corporate management or other resumes do not appear. All the material can be arranged with specific pieces for individual clients, and each piece was designed to stand alone.

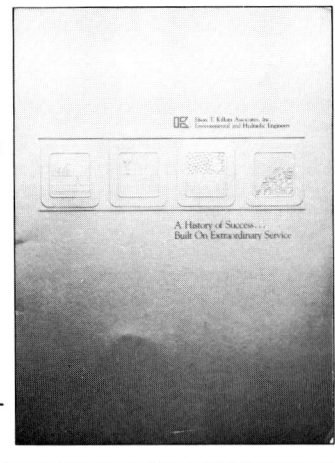

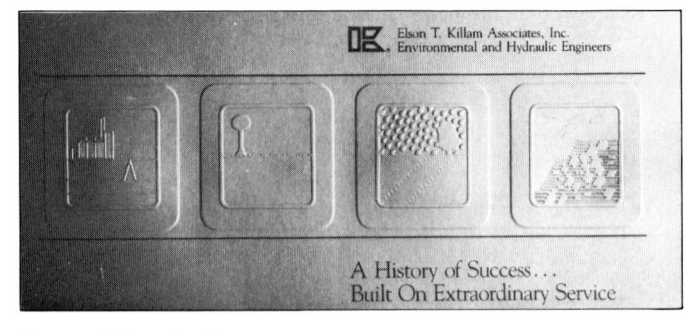

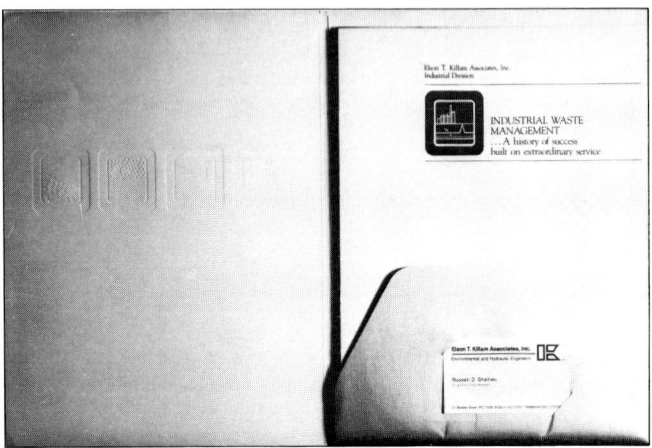

Presentation Folders

Presentation folders are silver with black and electric-blue type, a green logo, and black rule lines. Embossed graphic symbols of specialty areas appear across the top of the piece. Inside is a pocket to hold specialty brochures and project pages.

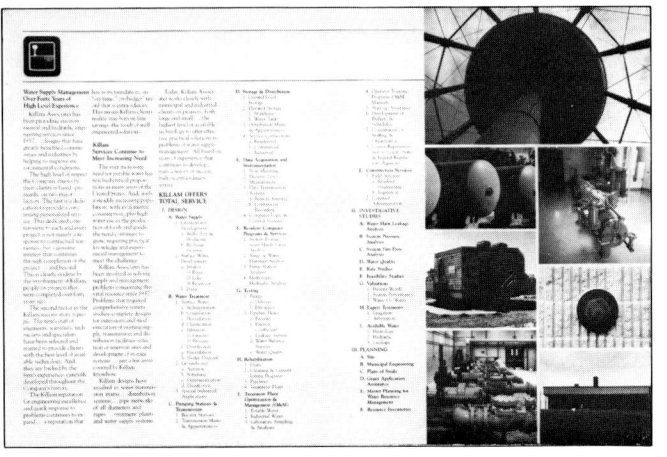

Company Brochure

Black type appears with embossed visuals across the top of the brochure cover, showing the graphic symbols for each specialty. Full-page color pictures show people working at project sites. The text describes the company's years of experience and its expertise in specialty areas.

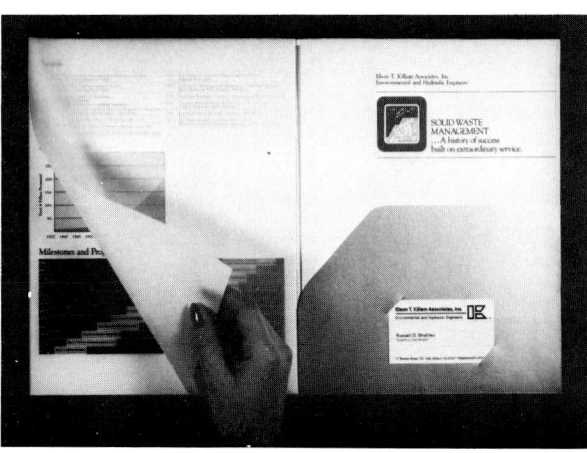

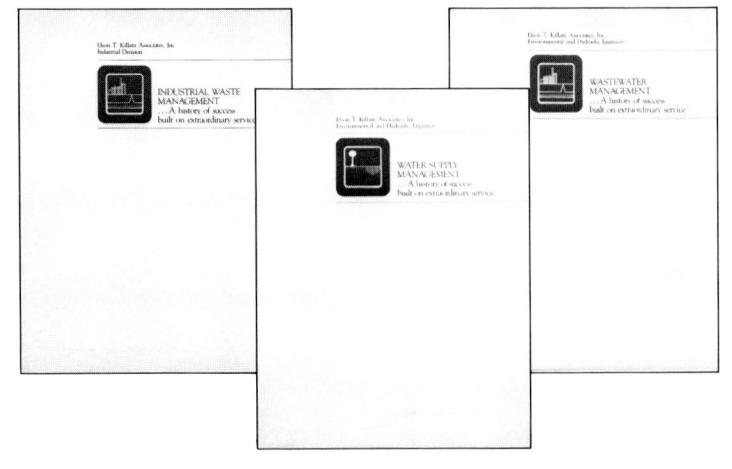

Van Dell

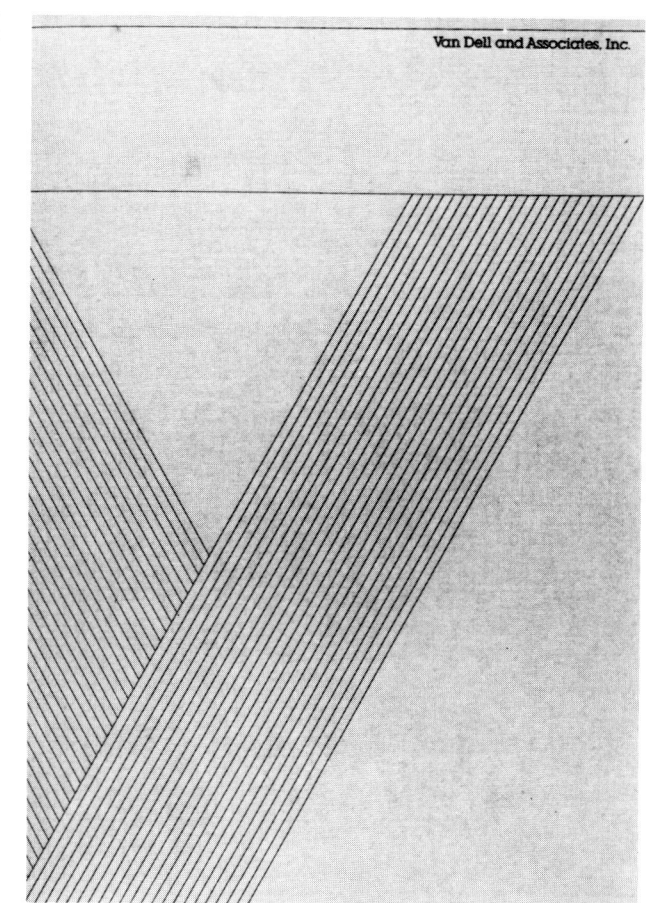

Company Brochure

The gray cover has dark-gray graphics and type with black rule lines. Inside, the firm's name appears in black. The brochure uses a four-column format. Visual symbols representing different aspects of the company's expertise appear across the top of the page. These illustrations appear throughout the material, creating the "brochure system." The back cover has a pocket and an insert area where business cards can be kept.

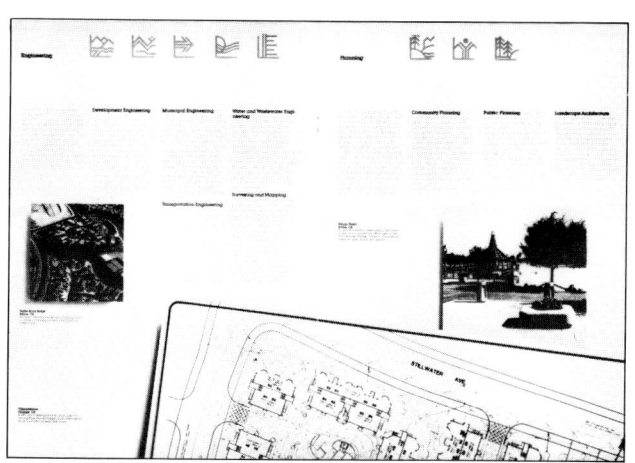

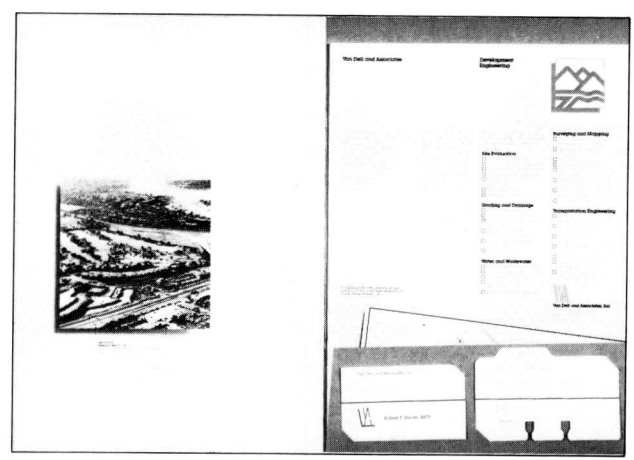

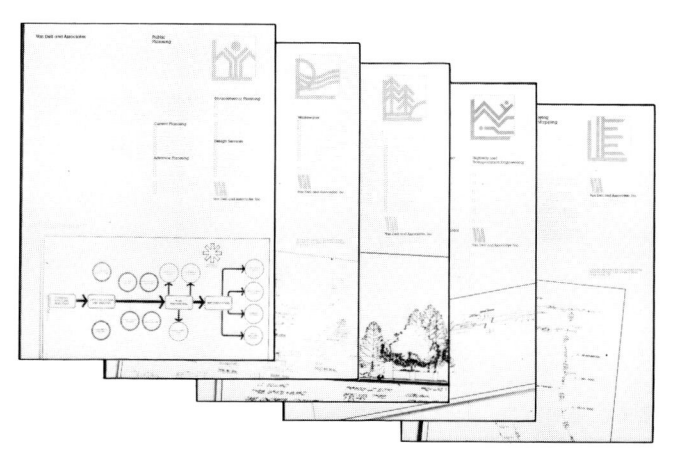

Project Pages

Black, white, and gray were chosen for these pages. The text discusses the company's history and areas of expertise. Keyed visuals appear at the top of the page. Renderings provide details of the projects and processes involved in the company's work.

Company Brochures: Engineers 135

The Earth Technology Corporation

This engineering firm uses a brochure system to describe all its services and areas of expertise. The use of several individual brochures allows the firm to discuss, in detail, the different specialty areas. The covers of the specialty brochures are designed with different colors (i.e., blue, red, and yellow), while the company's main brochure is beige. The specialty brochures all carry a striped design across the top of the cover where the title appears. A cropped four-color photo is also used on the cover of the specialty brochures. Each of these photos is bordered in white and represents a particular area of expertise. These brochures all carry the company's logo.

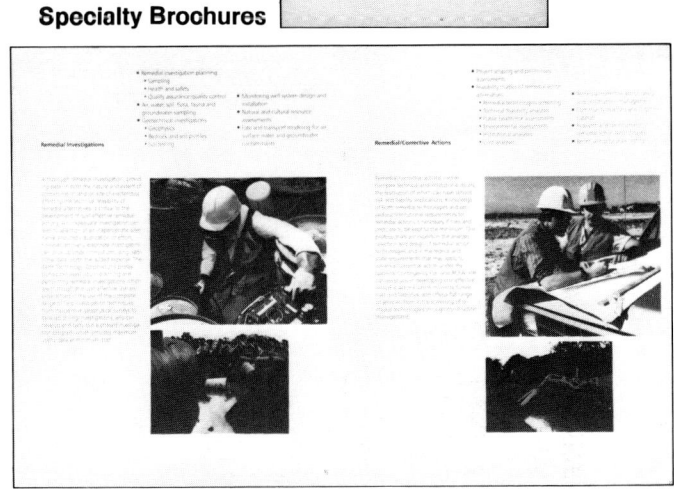

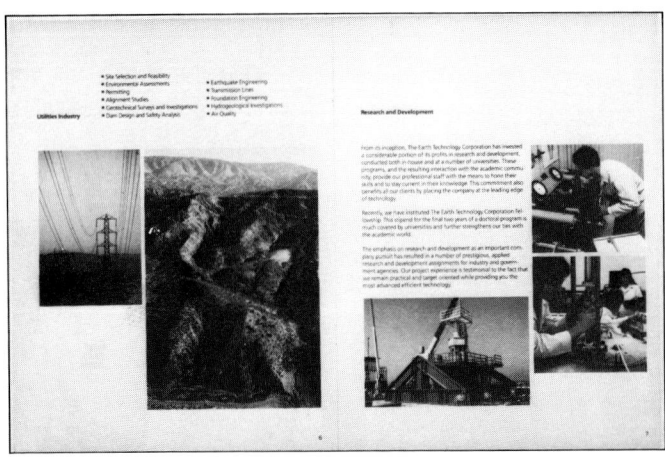

Specialty Brochures

Company Brochure

Using people in the photos is an integral part of the brochure, since the company wanted to show that although it is technically oriented, the human factor is imperative to its success. The last page lists the firm's locations.

Rocky Mountain Geotechnical

This environmental engineering firm uses several direct-mail pieces with its marketing materials. This piece, which fits into a standard size envelope, opens to a three-fold. The cover consists of the firm's logo and states the purpose of the piece.

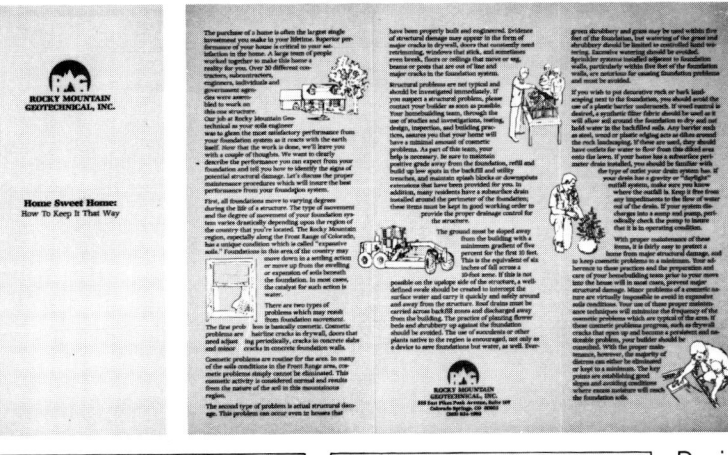

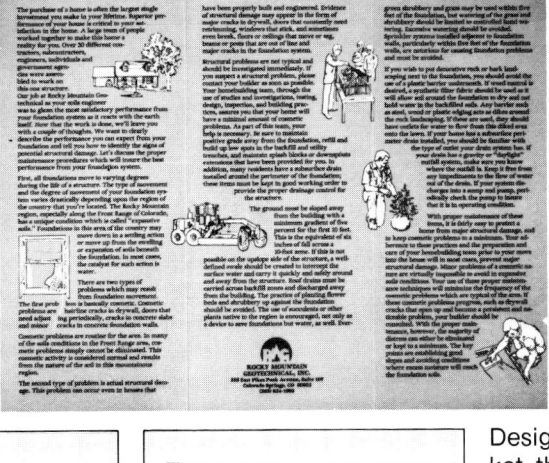

Direct Mail

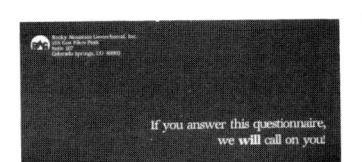

Specialty Brochure

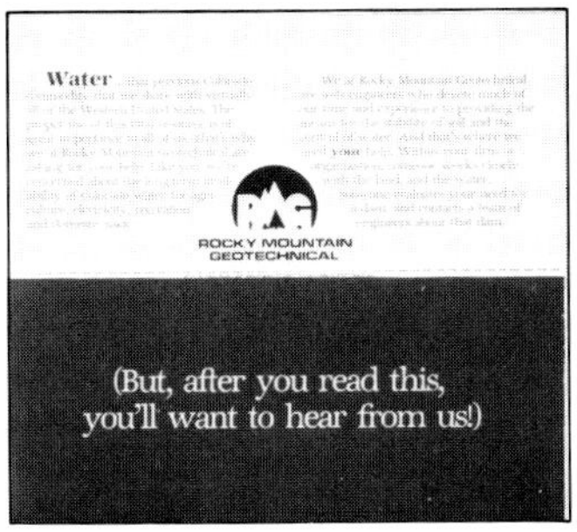

Designed for a special market, this horizontally-held piece captures the reader's attention by making two very positive statements. The inside contains a brief amount of text and a questionnaire. After filling out the questionnaire, the firm invites the reader to respond for more information by simply filling in their names and mailing in a supplied self addressed, stamped panel.

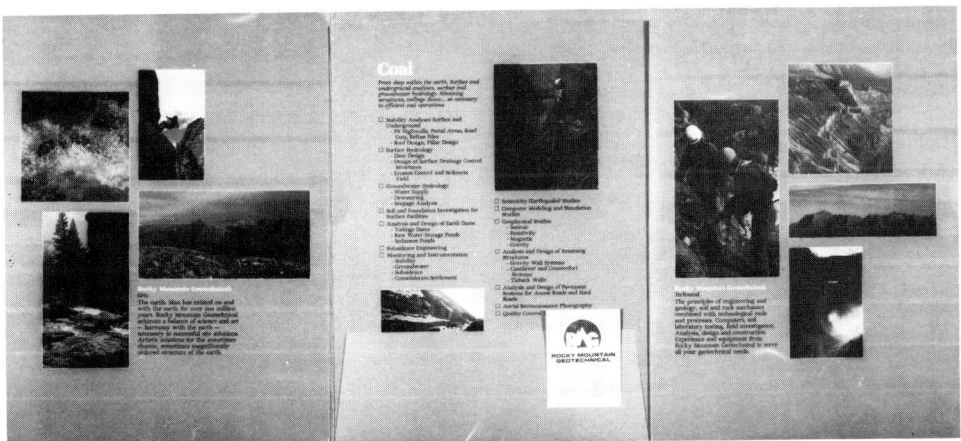

Company Brochure

This brochure functions as a main marketing piece and a response to direct mail requests. As a combination brochure and folder, this three-fold piece exhibits several full-color photographs of the unspoiled wilderness. The center pocket holds several full-color single pages, separately listing each of the firm's areas of expertise. A check list of the capabilities in that specific area is also given.

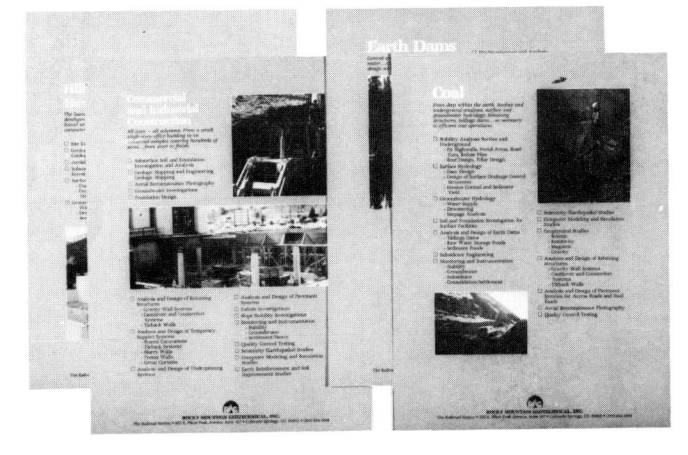

Company Brochures: Engineers 137

M.J. Brock

This construction firm has designed a brochure system to market its services. The system is based on a two-color cover scheme. Each specialty brochure has one maroon high-contrast photo on its cover, pertaining to the subject of the brochure.

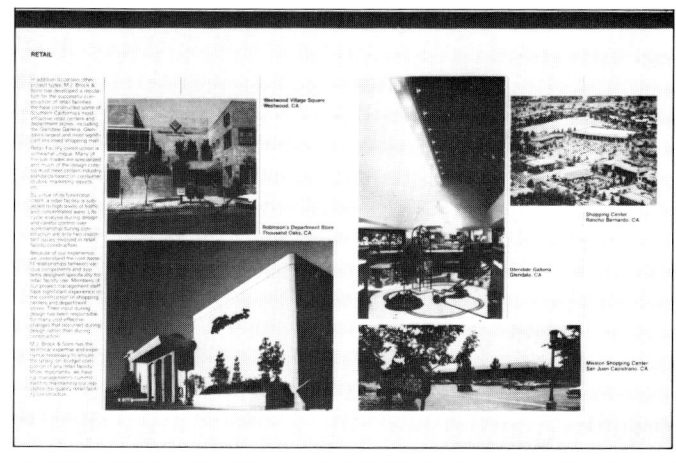

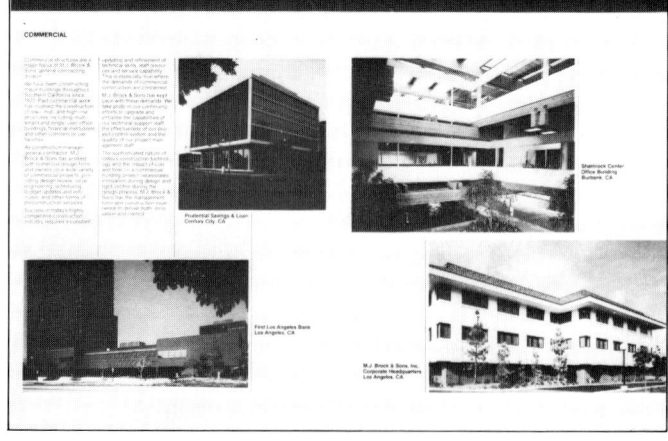

138

The main brochure has all of the specialty cover photos on its cover. These same one-color high-contrast photographs can also be found on the inside of each. Other full-color photographs are featured in these two page pieces, along with brief text. The pages are divided up by maroon or gray rule lines. The folder that holds these pieces has a simple gray cover with the name of the company in white drop-out type. A thick maroon border is featured at the top of each page.

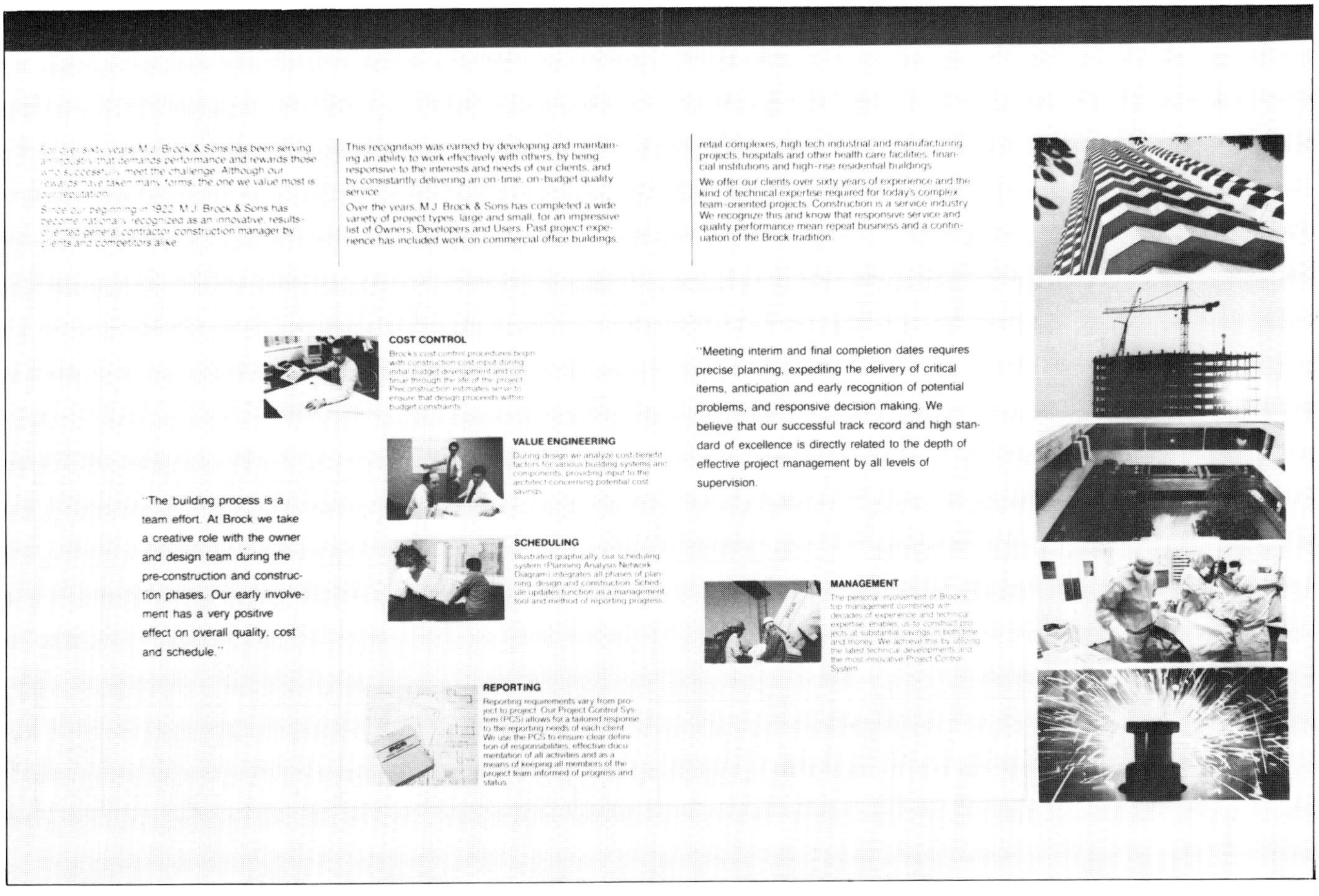

Company Brochures: Engineers

Cini-Grissom

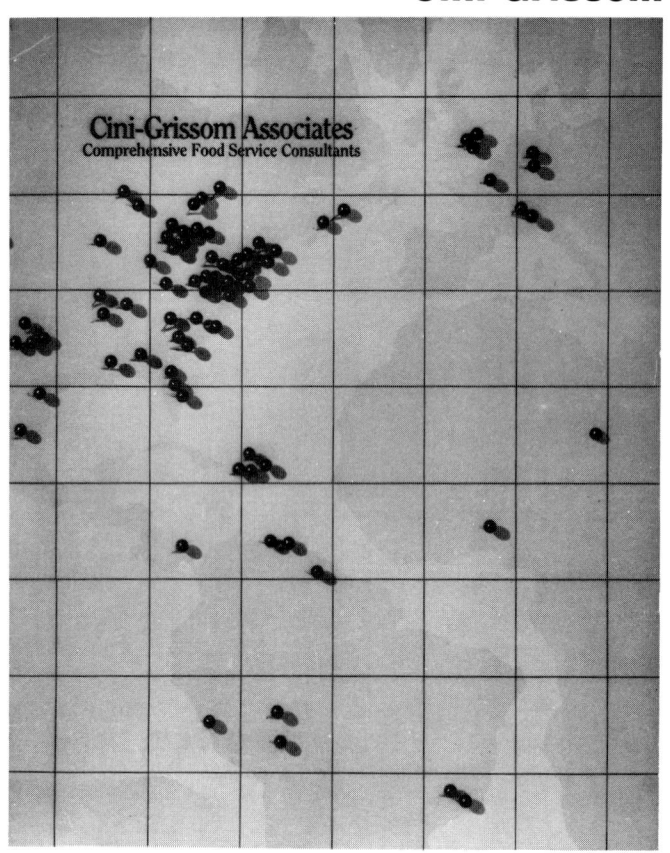

Specialty Pages

The two-sided pages are in the same beige as all of the other materials and show a varnished design of a map with pins (indicating where the firm has offices). One side has a black and white photograph. The text details the firm's involvements. At the bottom of the front side, the firm's address is given. The reverse side of the page gives a detailed list of projects. These pages are held in a folder with a grid pattern and a varnish of a map, also showing pins indicating the firm's locations.

Reprints

As part of its marketing materials, the firm also includes reprints of articles (which may appear in industry publications) on its services.

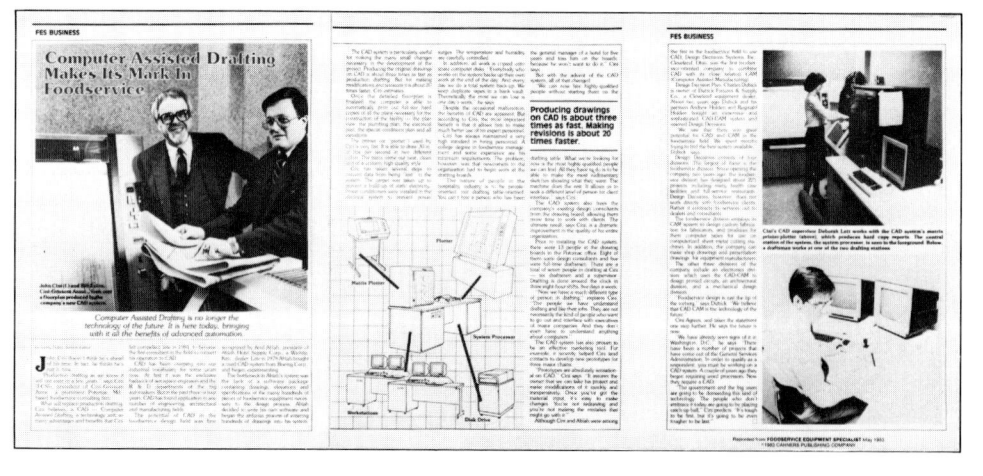

Newsletter

The company newsletter uses a three-column format. Black and white photographs are used as well as beige and black accents, including beige and black tab lines. The four-page piece discusses current trends in the food industry and furnishes other pertinent information. At times, the newsletter includes recipes, printed on a separate sheet.

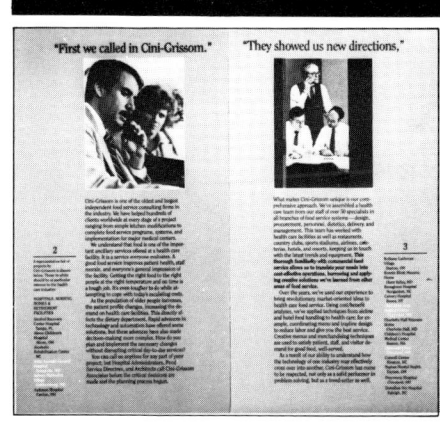

Resume Pages

A black and white photograph appears on the cover of this piece, along with client quotes that can also be found throughout the piece. Inside, black and white photographs show people in their workplace. These visuals are complemented by copy that relates directly to the photos. A partial client list is included on each page. Individual profiles of company executives fit into a pocket.

Case Studies: Consultants

Stratton

A brochure system is used by this general contracting firm. The material is stored in a sturdy 1-color folder. The folder is a gatefold, and provides 4 pockets inside. It folds compactly to be handled and carried, and includes spec folders, client list folders, newsletters, and reprints. The name of the firm appears in a blind emboss on the cover. When folded out once, the name is seen larger, in strong forest green letters. When it is folded out completely, the name is repeated, larger and off the sides of the piece.

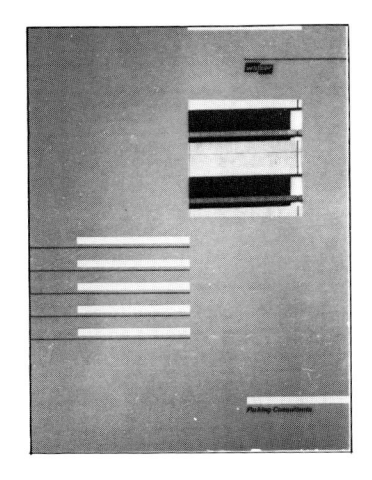

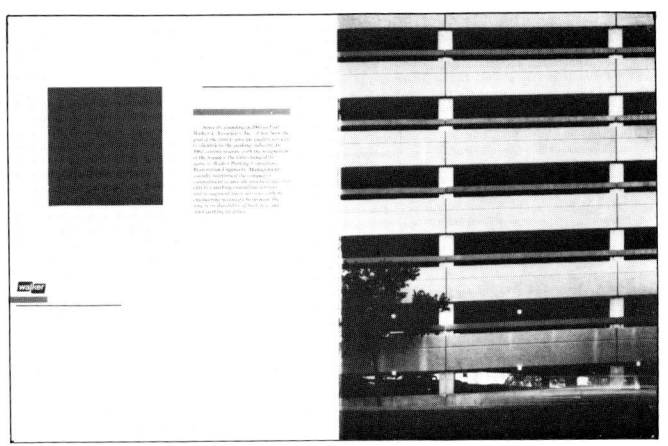

Walker

The targeted audiences for this firm's marketing materials include developers, governmental, hospital/medical, and educational clients. The two primary services provided by this firm are parking facilities consulting and restoration engineering. Separate brochures were created for each function, using one design to create a unified look. The colors selected complement each other while remaining neutral and understated. The firm's logo was created to resemble poured concrete and the cover employs a printing process which gives the graphic bars a gritty texture resembling concrete. The parking facilities brochure cover features a die-cut window exposing a portion of a parking deck photograph.

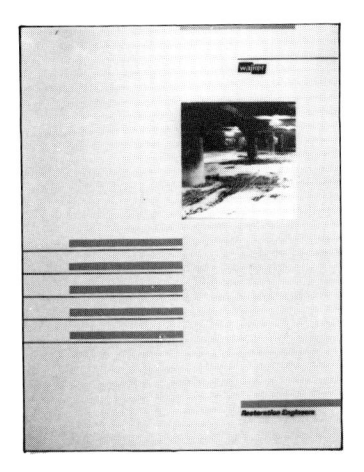

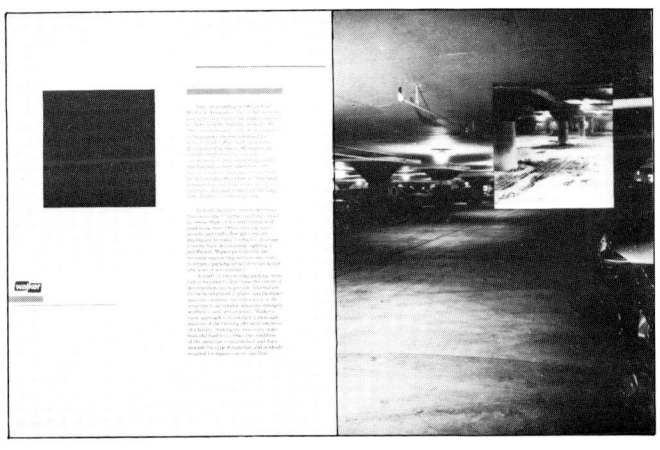

The restoration brochure cover features the same die-cut window as the parking facilities brochure. This brochure is a three-page foldout. The last page has a pocket where project pages can be stored.

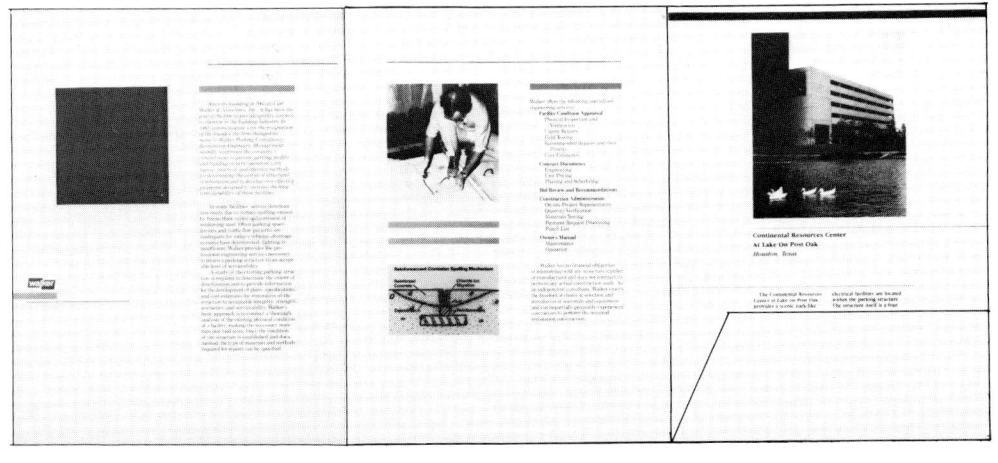

Company Brochures: Construction 143

Blount

This piece was designed to be read by chief executive officers and presidents of major corporations. The brief text reflects the concerns and philosophies of the firm, while colored renderings and striking photographs display the company's completed projects.

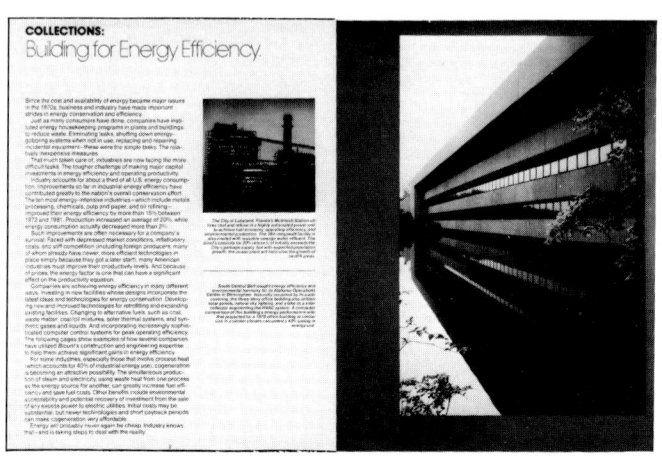

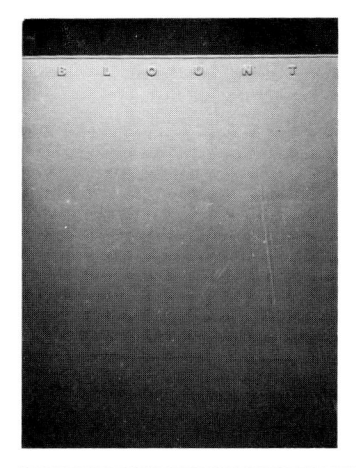

Company Brochure

This piece has been designed with a copper and maroon cover and embossed rule line and type. The inside cover is turquoise with white type. Dramatic views of company projects are shown in full-bleed color photos. The text is minimal, pertaining to each completed project.

Specialty Brochure

Black mat spot varnish and glossy black lines are used on this brochure cover. The same typeface and rule lines as the other pieces are maintained. Dramatic color photographs on a black field with white borders are used throughout the piece. Glossy black rules on a mat black field appear on every page. The text appears in a three-column format with red headlines. The text discusses the firm's specialty areas.

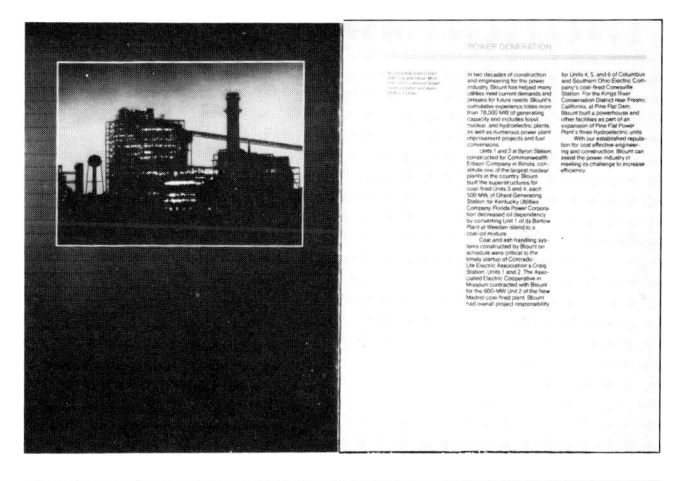

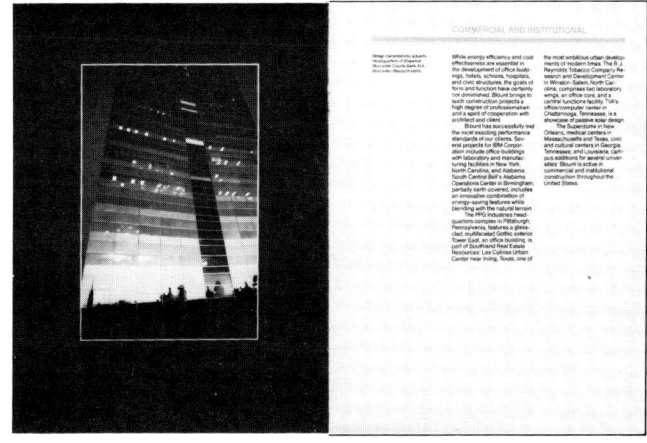

Company Brochures: Engineering 145

Representative Projects

Sverdrup provides engineering, architecture, planning, construction, and project management services in the United States and abroad. We've built a reputation for skillful handling of major projects — and for close, careful attention to the details of even the smallest jobs.

The laboratory—
where discovery becomes a business.

How to improve laboratory productivity, energy-efficiency, flexibility, safety, and cost-effectiveness.

NEWS OF THE SVERDRUP COMPANIES & PEOPLE

SPICE
TECHNOLOGY

Navy's Largest Job Begins

Sverdrup Architecture

Sverdrup Leaders for your airport design and construction team.

How STL built around 12,007,363 passengers

Growing airport maintains operations while doubling capacity

CASE STUDIES: 6

Vitetta Group

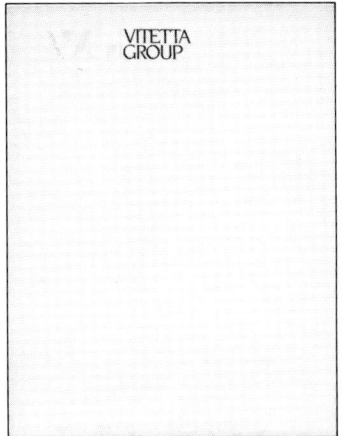

This glossy white heavy-stock folder features a gray logo and brown and gray type. The name of the firm appears down the false binding. The folder opens to reveal two pockets where material such as project pages can be held. There is also a place for a business card.

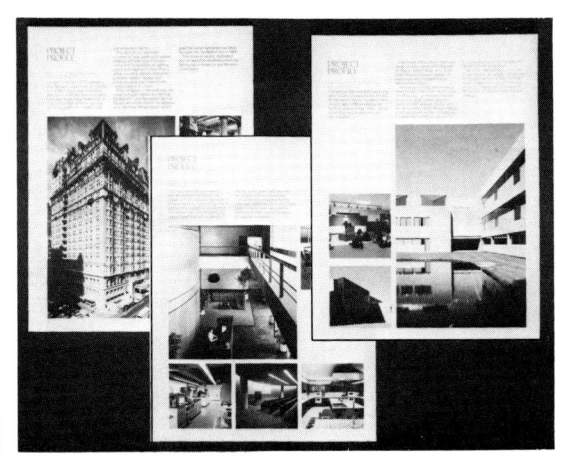

Project Pages
One-sided project pages with text and color photographs show completed projects.

Newsletter
The quarterly newsletter runs four pages and includes notes on the firm's projects and information about the staff. The two-color pieces use small project photographs. A thick black rule line across the top half of the page is repeated on every page.

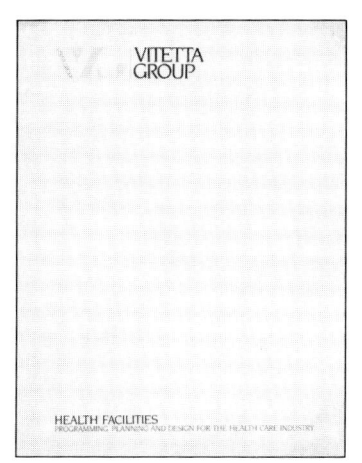

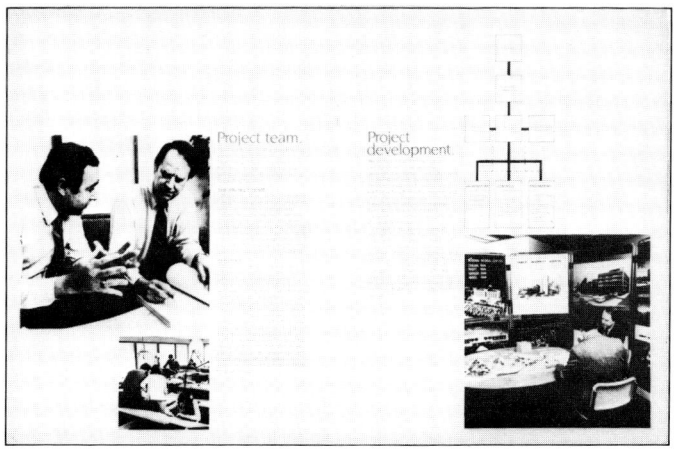

Specialty Brochure

These brochures have the same cover as the main folder. They are designed to highlight the firm's special areas of expertise. Photos of staff members are shown and full-page photographs of projects are included. The back covers of these individual brochures carry a specific one-line slogan.

Company Brochure

The cover of this piece is like the cover for the firm's specialty brochures, except that it lists the disciplines. Inside, bold black and white photographs accompany the text. On some pages there are one or more visuals, and on the others, text is used alone. Some pages also show people working with each other or using computers.

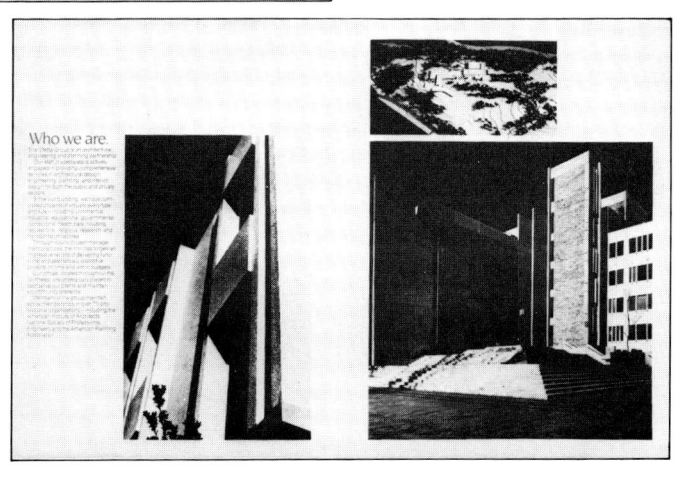

Case Studies: Architects 149

Catalyst

The marketing materials created by this architectural firm are graphically tied together by the use of a consistently applied corporate identity program. A distinctive logo and logotype create a strong link between the name of the firm and its services through this consistent application of its graphic materials

Direct Mail

This direct mail piece uses a black-and-white photograph on its cover and makes a provocative statement which entices the reader to open the piece and read on. The inside of this flyer is printed in full, vibrant color that visually jumps out at the reader. The answer to the opening statement is shown inside along with text on the firms capabilities and experience.

Project Update Newsletter

A unique folder format was chosen by this firm for its quarterly project newsletter. The outside of the folder is an illustrated grid of glass-blocks and tile. When it is opened, black-and-white photographs and text (which discusses company updates, awards, publications, services and recently completed project information) is revealed. Other single sheets, which pertain to specialized areas of architecture, can be inserted into this folder.

Newsletter

The monthly newsletter created by this firm also uses a unique format. The piece, which is designed to be a self-mailer, folds out into six panels. All text, photographs, and renderings sit in ruled boxes which, by the use of a side shadow, seem to be raised above the light gray grid background. The information which covers all aspects of the firm's activities is printed on both sides of these panels. The piece can be folded out once again to reveal a three-color poster that shows detail photographs of projects arranged abstractly on a black-and-white grid background.

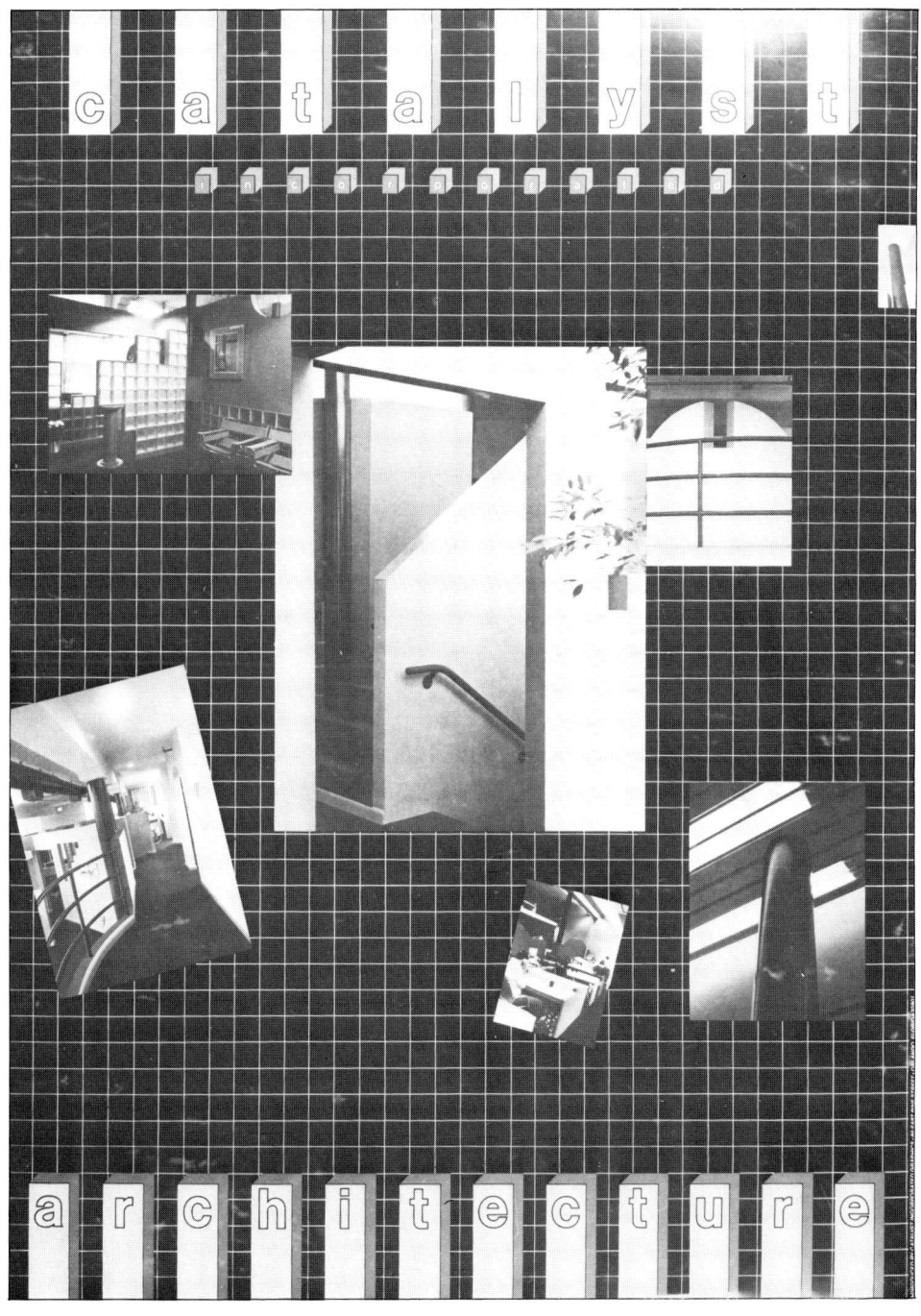

Case Studies: Architects 151

Zimmer Gunsul Frasca Partnership

A blue folder with a die-cut and gold lettering opens to reveal two scalloped pockets. The pockets hold project pages, reprints of articles, a resume brochure, and an introductory company brochure. The same die-cut used on the front cover is also used on the two inside pockets and on the back cover.

Reprints

The reprints used by this firm are the same format as the magazine from which they came and the type and photographs are lifted directly from its pages. All reprints are used in full color.

152

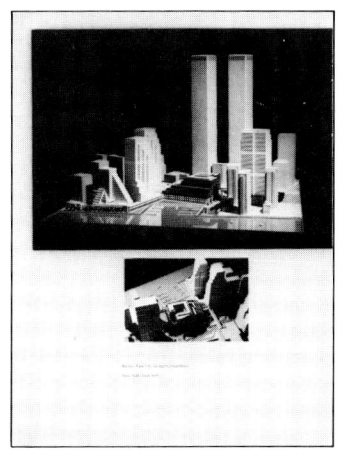

Project Pages

These pages inform prospective clients of the type of services the firm provides. This marketing tool features color photographs of projects and aerial view renderings to illustrate the firm's expertise.

Resume Brochure

This brochure contains information about the firm, photographs of employees, a brief resume on these people, a list of publications about completed projects, and an awards list.

Company Brochure

The almond-colored cover carries the name of the firm and the company logo. Inside, photographs show firm members working. Brief text accompanies many of the photographs.

Case Studies: Architects 153

THE CALLISON PARTNERSHIP

These four pieces were developed to inform clients about specific company services. Individual silver-gray flyers are used to emphasize each discipline. All the flyers include the company's name, address, and telephone number.

Company Brochure

The silver cover features a color photograph. A tissue overlay leads to a page of quotes from the company executives. The brochure includes minimal text, just enough to inform readers without overloading them. Color photographs of completed projects, interiors, and detail photographs illustrate the company's ongoing goals and accomplishments.

Quarterly Newsletter

The newsletter informs clients of the firm's current activities and capabilities. The design depends on instant graphic appeal. The text is brief, supplemented with pictures to maintain interest.

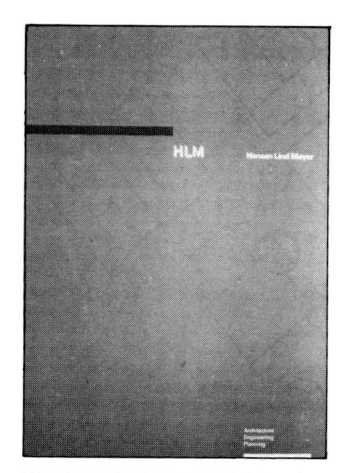

HLM

This gray piece shows a varnished imprint of a floor plan, which is repeated in parts throughout the brochure. The front cover has white type and blue and white tab lines. The text discusses the firms strong commitment to its field and its approach to its work. Four-color photos are used on every page. The text emphasizes the solutions the firm is offering the client. Aerial and wide-angle views are used generously, giving the piece a bold look. Interiors and exteriors are shown, as well as renderings and models.

Folder

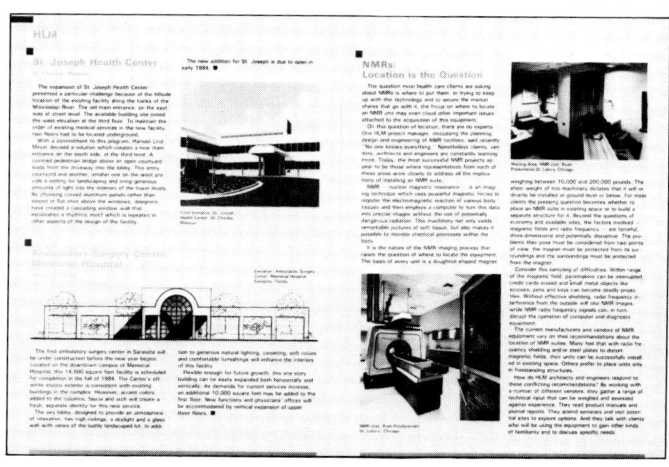

Newsletter

The graphic style used on all of this firm's promotional materials carries through to its newsletter. Though the piece is printed primarily in black-and-white, the firm chose to use blue graphic bars and squares to highlight each article. These touches of blue visually relate this newsletter to the firm's other pieces.

Case Studies: Architects 155

Perkins & Will

This quarterly newsletter uses burgundy type and rule lines throughout its pages. The layout employs three columns of text and a grid-like pattern. Articles include a report style analysis of the firm's work and discussions of ongoing and future projects.

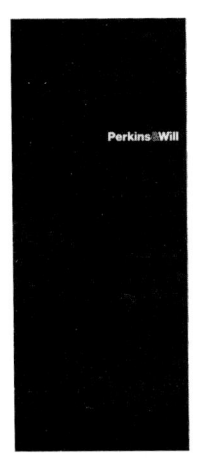

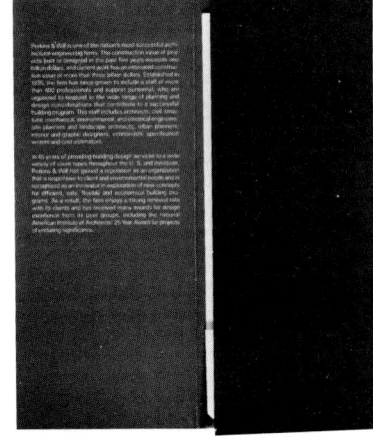

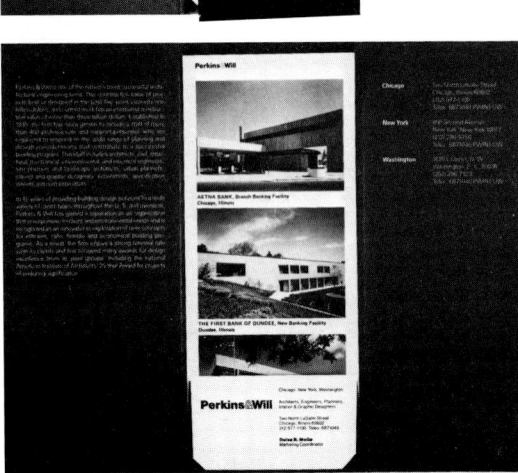

Folder and Newsletter

Direct Mail

The firm has also created direct mail brochures. These are envelope-sized pieces that fold out into three pages. Inside, mini-project pages in black and white are held in a pocket, along with client lists and other information about the firm.

156

Project Pages

Project pages are two-sided. One side pictures a color visual with text. The reverse side shows details of the site in black and white. Renderings are also used. Visuals laid out in a grid formation are featured on these pages.

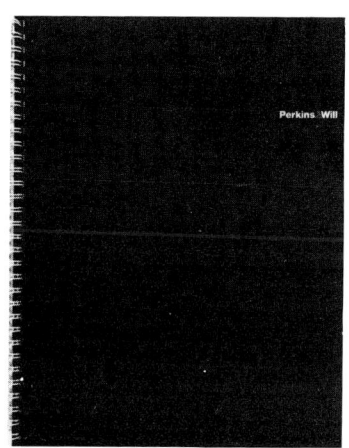

Company Brochure

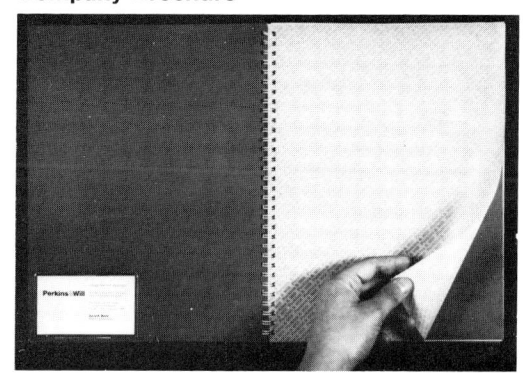

This wire spiral-bound brochure is composed of individual project pages, and has a plastic pocket, which holds a business card, on the inside cover. A frosted vellum overlay, with a pattern made up of the firm's name printed diagonally, introduces the reader to the piece. A similar overlay is repeated at the end of the brochure.

Case Studies: A / E 157

Gresham, Smith and Partners

Corporate Identity

A folder designed with an interior envelope and die-cuts to hold a business card can hold a combination of materials such as resumes, newsletters and specialty brochures. The corporate stationery uses the same prominent logo as seen on all the firm's promotional pieces.

Newsletter

This monthly twelve-page, black-and-white newsletter features information about the people in the firm and the firm's projects. The text includes information about the firm's goals and history. The newsletter's design differs from the company's other material by use of a red rule, which is maintained throughout the piece.

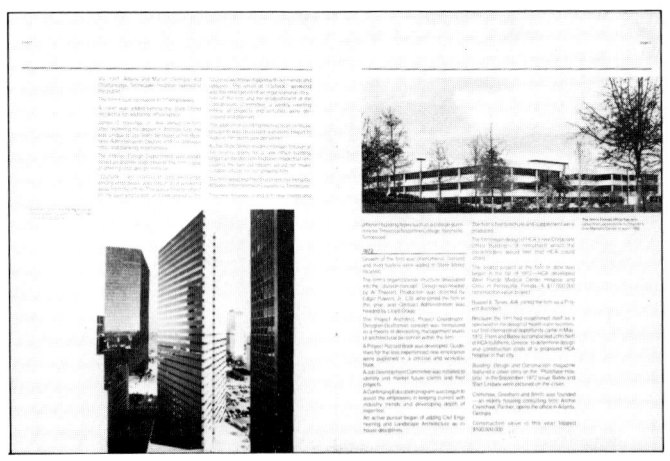

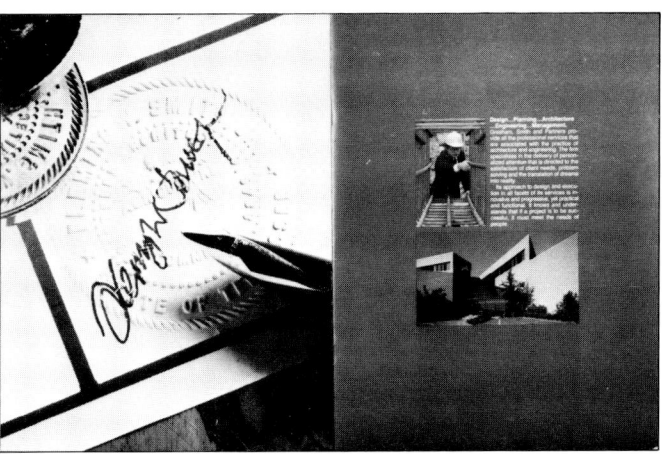

Company Brochure

The main company brochure carries the same silver logo printed on a heavy gray stock. Inside, full-bleed color photographs are used in conjunction with smaller photographs. All photographs are set in a solid gray background. The minimal text is dropped-out of the background. The text describes the firm's services and its approach to its work.

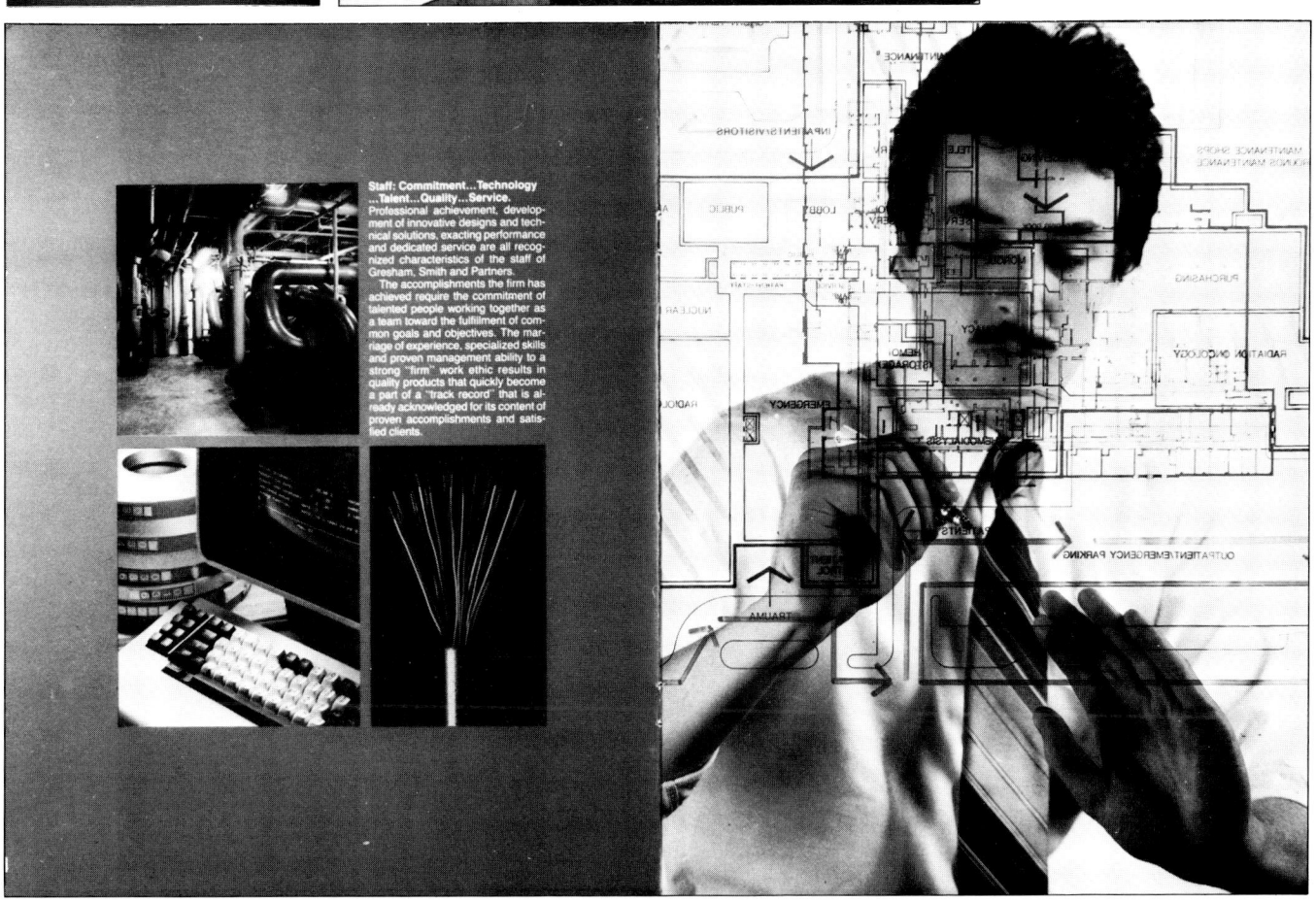

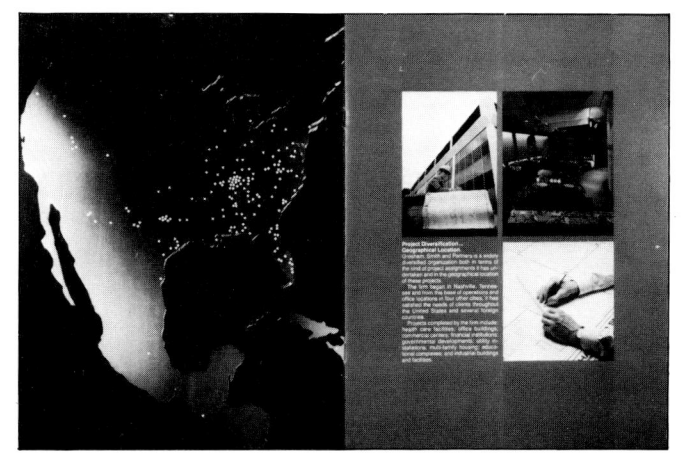

Case Studies: A / E   159

Gensler

Project Pages

The firm has over ninety-three fact sheets of interior and exterior projects. These sheets picture the project, in color, on the lower half of the page. Usually, a black-and-white rendering or a floor plan is included. On the top portion of the sheet, text describes the project.

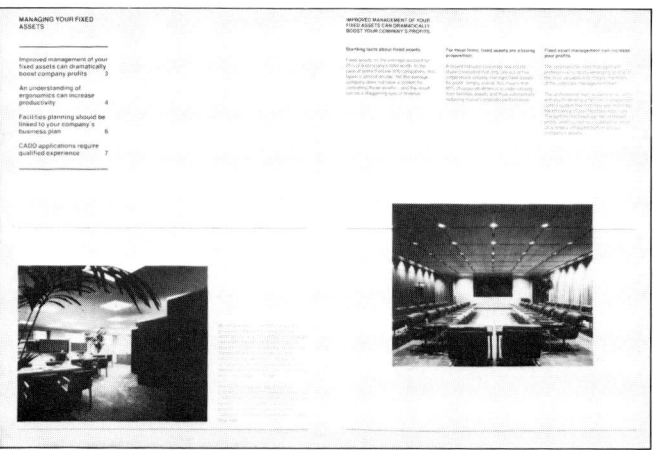

Newsletter

This quarterly newsletter focuses on services, and updates recent projects. Half of the pieces are printed and mailed, and the other half are printed without the date so that they can be used in general qualification packages. The text is minimal, usually discussing the project or where the firm is in terms of growth.

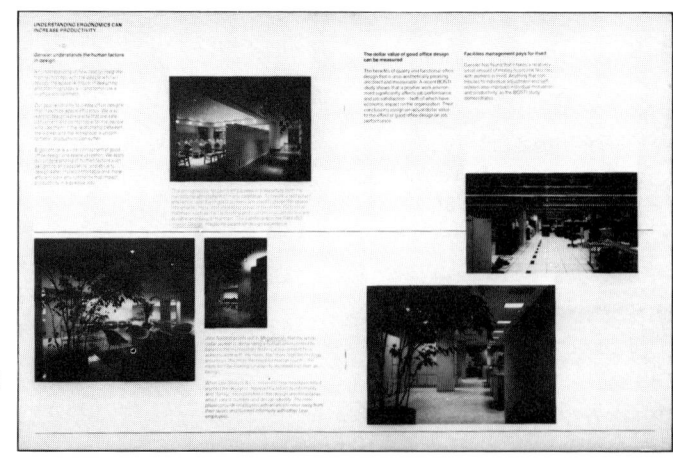

Specialty Brochures

These specialty brochures resemble the quarterly newsletter. Inside, black-and-white photographs are used, across the top of the page. Text at the top of the page describes the firm's expertise in a specific area.

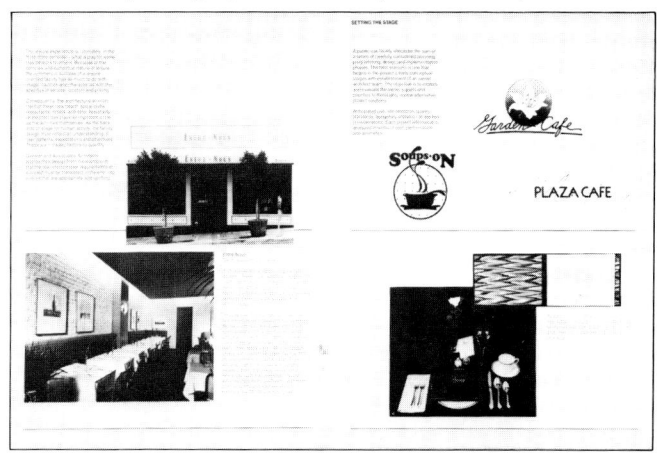

Regional Office Brochure

The gray cover is simple, showing only white type (the firm's name) and a color photograph. Inside, color photographs of interiors and exteriors are shown. The text is brief, and discusses the regional architectural design office.

CADD Pocket Brochure

This brochure was designed to market the firm's computer-aided drafting and design services. Computer-assisted drafting and design is a system which the firm wanted to present to the reader in an unintimidating manner. The bright yellow cover is "friendly," and the inside piece shows playful cartoons on white pages. The straightfoward text explains the system.

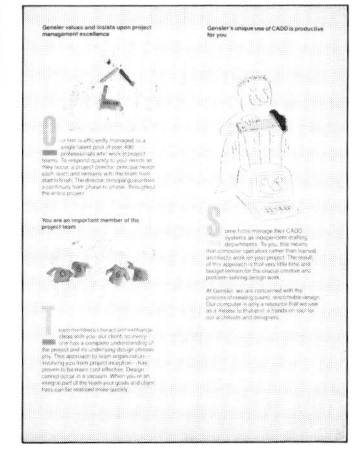

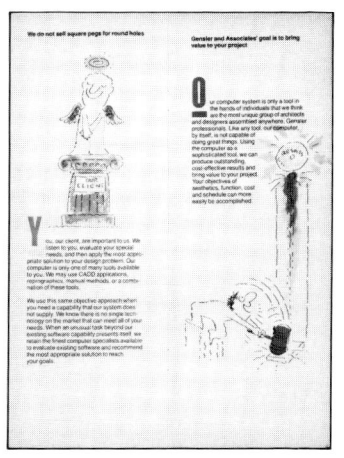

Case Studies: Architects 161

Thompson Ventulett Stainback

Folder

A glossy lacquered white package holds various company material in scallop-shaped pockets.

Discipline Brochure

This mixed-use design brochure is geared toward commercial developers, hotel operators, and public authorities. The hotel design brochure is aimed at hotel operators and developers. Other special market brochures include the retail design brochure and office design brochure. Each brochure is used to establish or reconfirm audience confidence and to keep clients in other markets informed of the firm's activities. These brochures can also be used in formal proposals combined with the firm's general capabilities brochure, with project photographs, and the proposal itself.

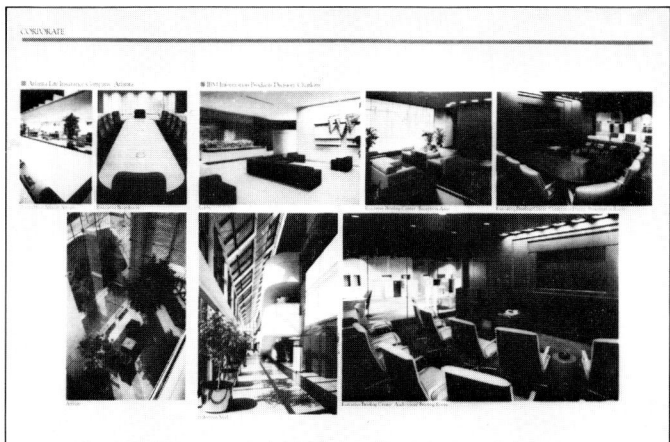

Discipline Brochure

The new interior design brochure is intended for corporate office builders, developers of office buildings, hotel operators, and builders. It is designed to be used when offering interior design services only, or it can be adapted to present the company's comprehensive architectural and design services.

Company Brochure

This brochure is targeted toward corporate facilities executives, commercial office building developers, hotel operators, and retailers. It is created to quickly offer an impression of the firm's capabilities, aided by color photographs which help establish this goal. The brochure is designed to be scanned and then read.

Short statements of the firm's client orientation caption many of the photo spreads. The text supplies the client with detailed information about the firm. This description includes a list of key clients, award-winning projects, and statements about how the firm is organized and how individual projects are implemented.

Case Studies: Architects 163

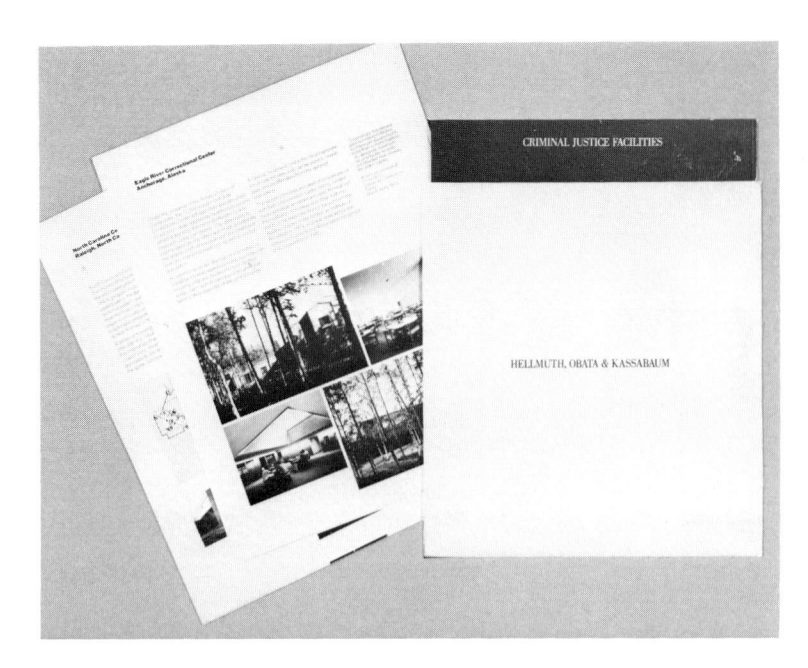

Hellmuth, Obata & Kassabaum

This architectural firm's marketing materials are designed as a completely flexible system. The folders that hold all of the firm's promotional materials can be tailored to the individual client also, by the use of tipped-in photographs on the covers. The first piece shown here is a pocket-style sleeve which can hold one of the firm's specialty brochures and selected project pages. This package can act as a direct mail introductory piece for targeted clients.

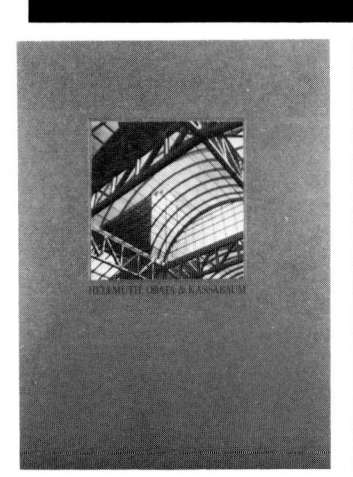

Project Pages/Folder

A cool-gray, textured stock is used for this folder. An embossed square acts as a frame for the selected tipped-in project photograph. The name for the firm is printed below this box. The format of the project pages varies from one-sided, full-color pieces to two-sided, black-and-white pages. Though all display photographs, some pages also use floor plans and renderings. The use of client quotes help make this extensive series of project pages unique and effective.

Folder

The folder for these brochures uses the same format and paper stock as the project page folder.

Specialty Brochures

The specialty brochures, which use a gate-fold format, are printed in different values of blue, green, and purple. The linear screened photographs used on the covers of these brochures is in the same position as those tipped-in on the folder covers. When opened to the full four-fold width, an array of full-color, keyed photographs of projects is revealed. The text is kept brief.

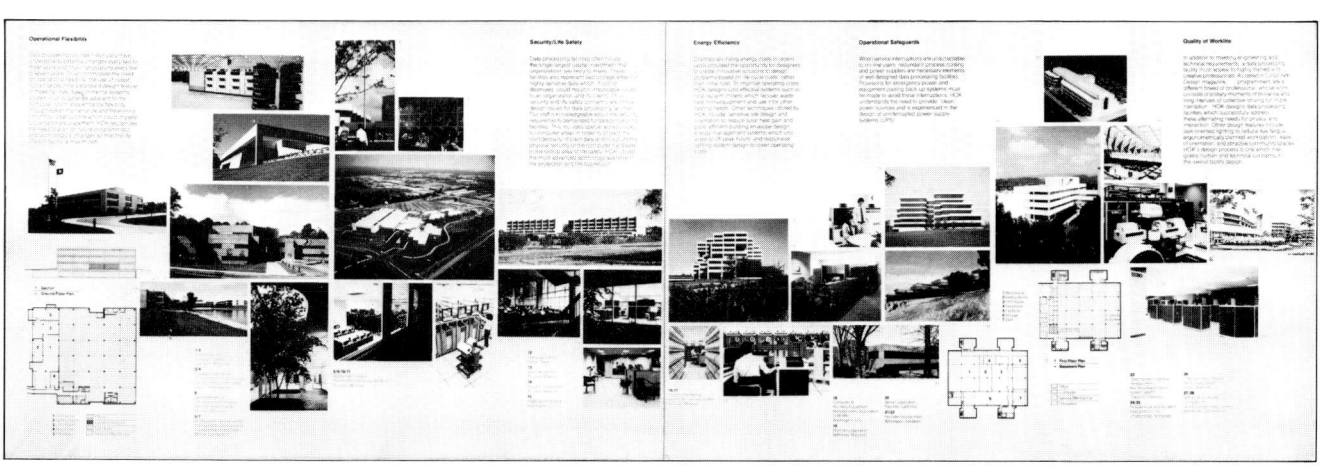

Case Studies: Architects 165

Swanke Hayden Connell

Quarterly Newsletter

A full-color newsletter with strong graphic ties to all other company promotional pieces was created by this architectural firm. The newsletter entitled "In House" gives several photographic views of recently completed projects as well as the descriptive text and floor plans. A full-page bleed photograph of the featured project makes up the cover.

Folder/Project Pages

The strongest graphic tie used in all of these pieces is the selection and use of two basic colors, a deep blue gray and a contrasting terra cotta. These two colors, along with the use of a centered logo and the use of graphic bars, also create a unified look. This folder holds project pages as well as specialized client information. These full-color project pages vary from a two-page folder to a two-sided single sheet. All are printed on a slick, cream-colored stock. Bold photographs, descriptive text and floor plans are used throughout these project pages.

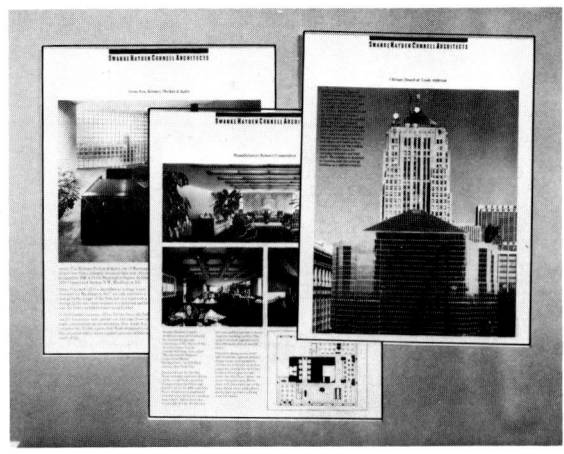

Company Brochure

In updating its company brochure, a main consideration was to try to keep the brochure "current" for as long as a ten-year period. This was accomplished by giving an overview of the firm without discussing specific projects. A full-color photograph was "tipped in" on the cover and can be changed to represent any one of three separate offices. The photograph is of a major project in that office's area (e.g., the cover photo used for the New York office is the Statue of Liberty restoration project.) Project summary sheets are created separately to tell the specific story of important projects and can be included in the brochure.

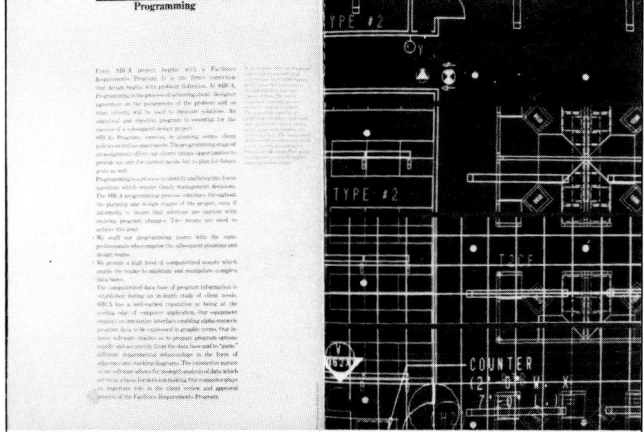

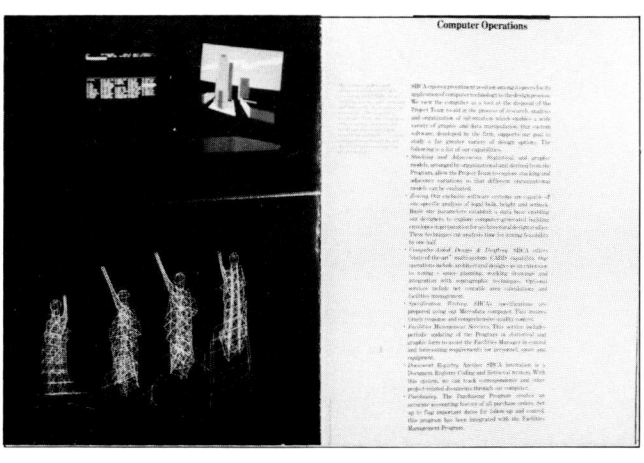

Case Studies: Architects 167

RTKL

This architectural and engineering firm designed a newsletter that opens up into a poster-size piece. This piece comes out quarterly and includes articles on current projects. Visuals include models, interiors, and renderings. The interior features black-and-white visuals as opposed to the outside of the piece, which is brightly colored. At times, four colors are used on the inside and an extra page is added. The newsletter is produced on folded 17x22-inch coated stock. The copy is brief and informative, with bold captions to catch the reader's attention. A special feature of this piece is the four-color poster on the back page which was conceived as an artistic expression with the firm's logo worked into the graphic.

Services Brochure

This architectural firm has created a brochure "system" consisting of individual marketing sheets covering a wide range of projects and services. The spiral binding allows the firm to create client-specific brochures quickly and efficiently. The lay-out of the pages follows a flexible yet consistent grid pattern. Project sheets are grouped under categories including retail, interiors, graphics, etc. Each section closes with a chronological listing of past and present projects.

Company Brochure

This brochure gives an overall view of this firm's many services. Full-color photographs are used on every page and most bleed across the gutter. The text (which is dropped-out in white against the matt-black backgrounds) is kept to a minimum, allowing the photography to tell the story of the firm's expertise. A partial client list is supplied at the bottom of every right-hand page.

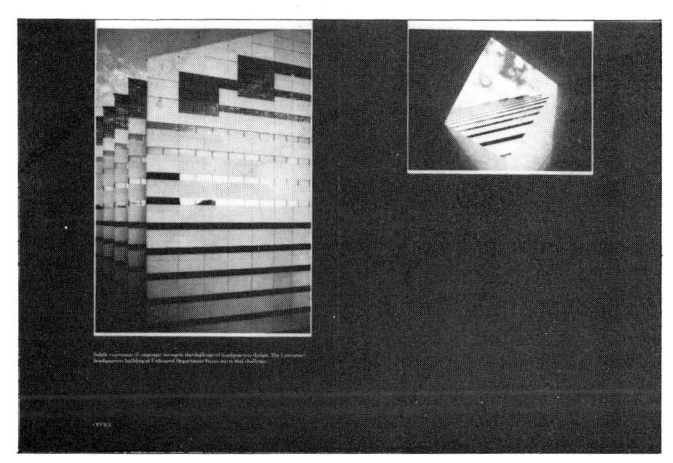

Case Studies: Architects 169

Brown And Caldwell

These two-sided, numbered pages are used to present the firm's experience in specific technical areas in a visually pleasing, readable, and persuasive manner. The project highlight pages briefly describe completed projects. These pages are gray and black duotone, with gray rules and logo echoing graphic elements which appear elsewhere in their proposal and marketing materials.

Client/Location Lists

The firm uses one-page client lists in its promotional material, as well as a one-page graphic of a map with office and laboratory locations highlighted.

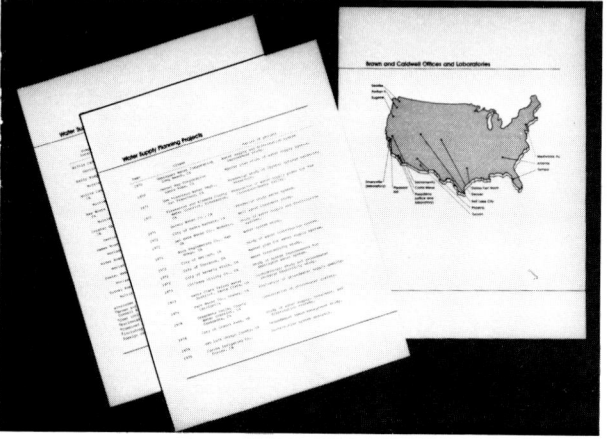

Quarterly Newsletter

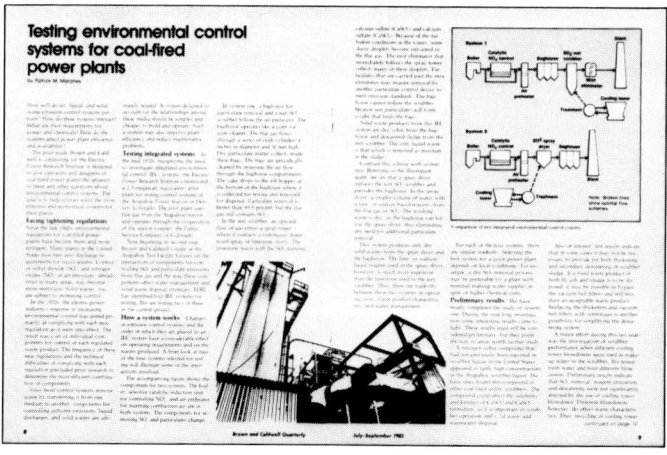

This publication is intended primarily for an external audience. Most of the material printed in this newsletter is of a technical nature, presented in an easy-to-understand format. The publication also gives recognition to the staff, and serves as an ongoing record of the firm's activities. Four-color visuals of recent projects are used on the cover, along with a brief description of what is contained in that issue.

Service Pages

This consulting engineering firm uses a number of marketing materials. Among these materials is a numbered series of two-sided black-and-white pages describing the firm's analytical services. The photos that are used in these pages are all company employees at work in the lab and the field.

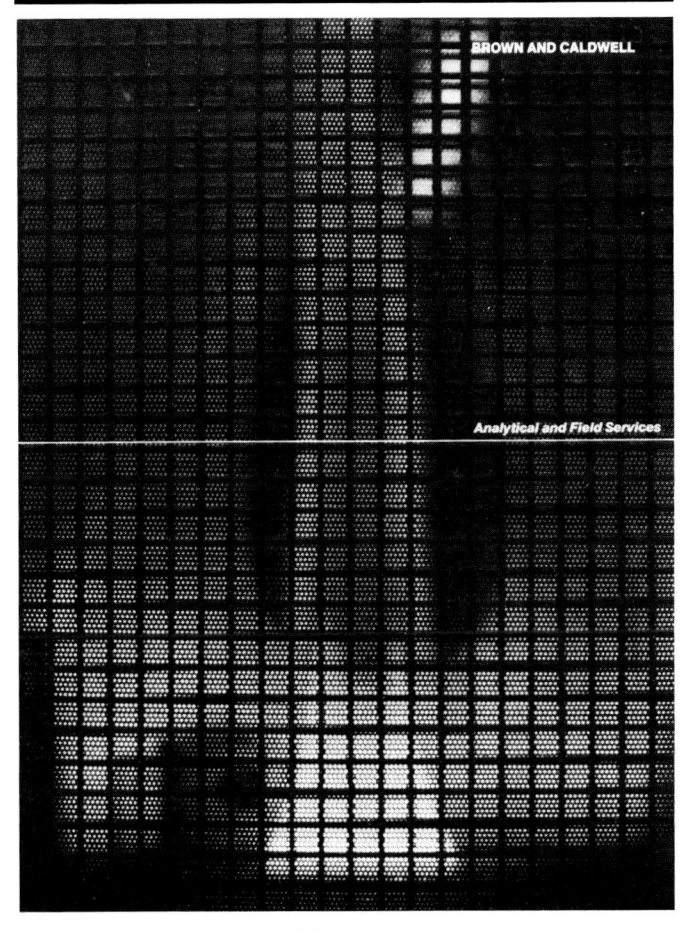

Specialty Brochures

These four-page specialty brochures use full page bleeds. The company name and the area of expertise, along with a thin rule, are used on the cover. The cover visual, as well as all of the inside photographs, are in full color. Black and gray type with black and gray rules are used. All of the photos on the inside of the brochure show people at work for the firm.

Case Studies: Engineers 171

Donohue

A vertical format was chosen which can be bound into proposals along with other material. The twelve-page format does not require a great deal of client effort to read. This piece fits into the overall corporate look by utilizing the narrow reverse rule design element. Four-color photographs lend additional visual impact.

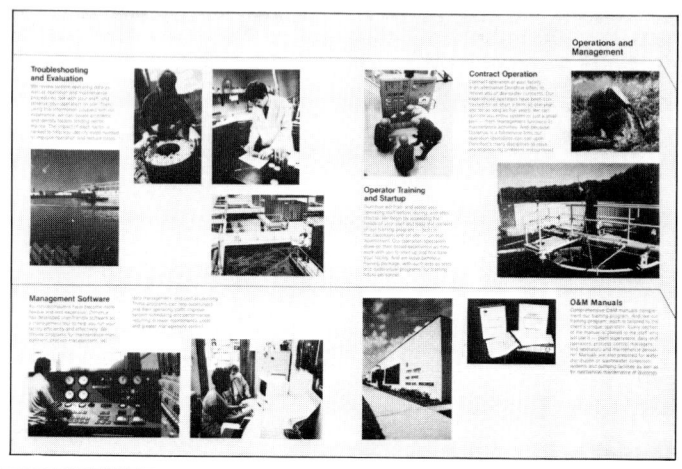

Magazine

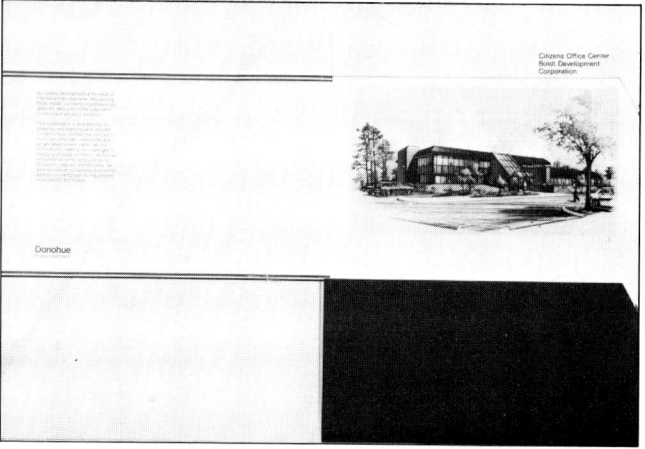

Project Pages

These well-rendered color pages with the firm's name, address, and telephone number are complete in themselves. Each location is described in detailed text to provide a project record which is useful to potential clients. Again, the narrow rules are continued on these pages.

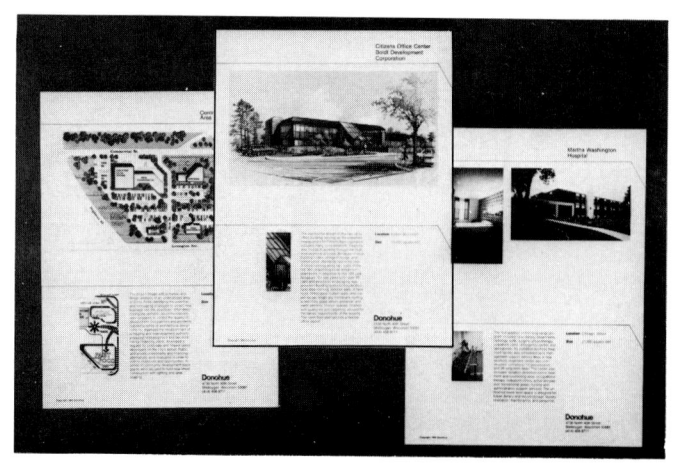

Anniversary Materials

Among the materials employed are holiday greeting cards, a 1985 calendar, a quarterly external publication, and announcements for open house functions and golf outings.

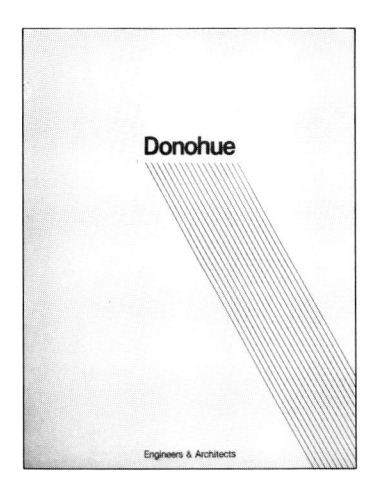

Company Brochure

Both a brochure and folder format were designed for this firm. The brochure is a stand-alone introductory piece which can be bound with other information in proposals. This large folder, with its inside flap, can hold specific service brochures, projects, information, corporate magazines, and reprints. The narrow rules on the cover are used throughout the brochure, including the project pages and individual brochures. Photo-collages are used for covers; the photographs reveal a range of services and represent a mix of projects. Minimum copy is used so that the piece does not become outdated.

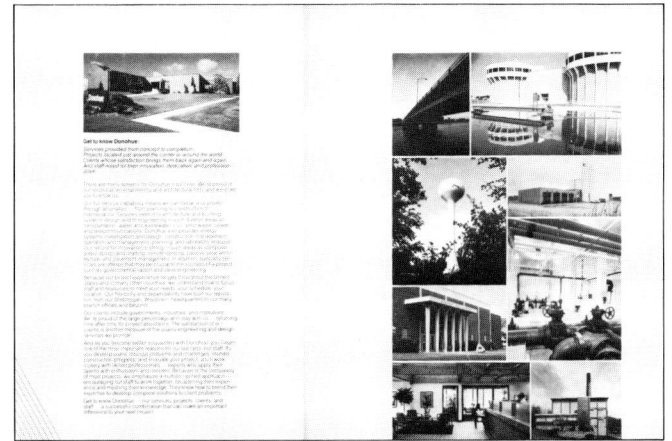

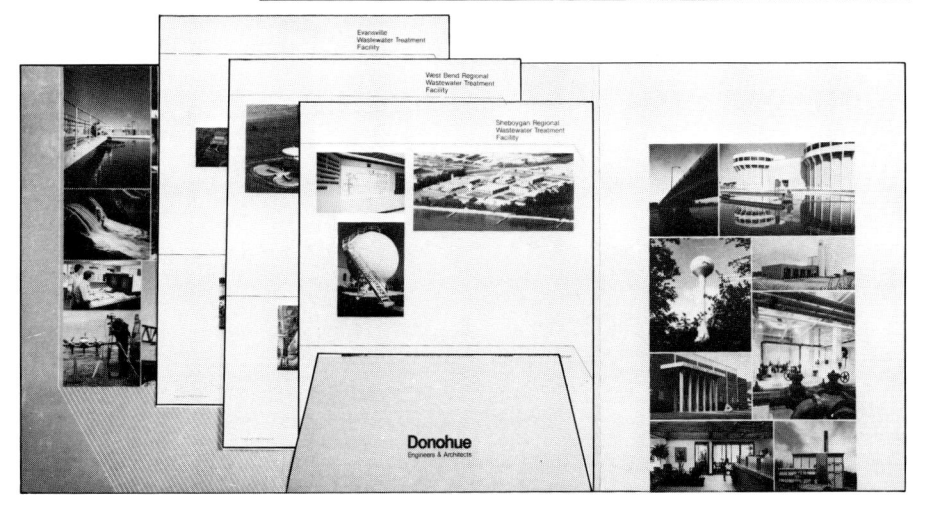

Case Studies: E / A

Syska & Hennessy

The four page newsletter, in two colors, is geared to specific regions. Two examples are the Cambridge version which has a green tab, and the New York version has a navy blue tab. The newsletters feature renderings and photographs of projects. Photographs of the employees are also shown. The text is kept brief, and is basically an update of the work in that region. A folder with a back pocket holds these pieces. This striking folder has a black cover with blue, green, and shades of maroon across the top. The name of the firm appears in white against the black field. Inside the folder is another statement from the chairman of the board.

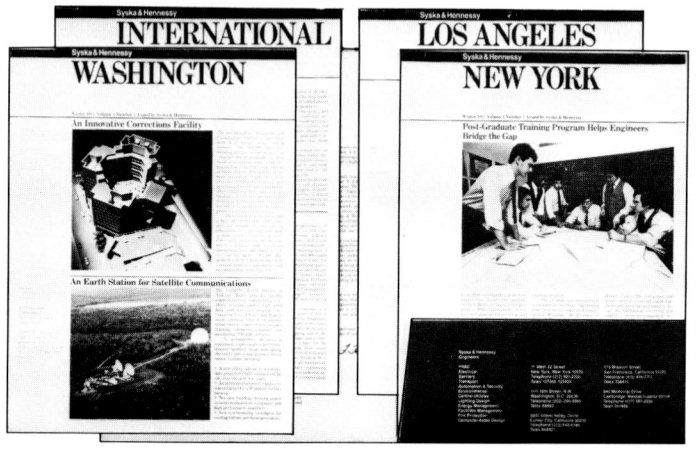

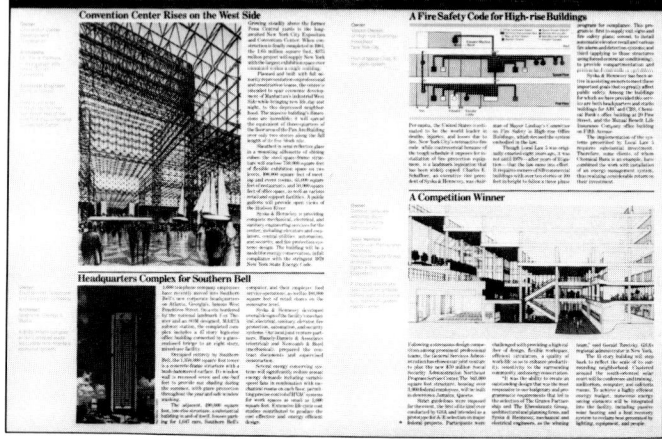

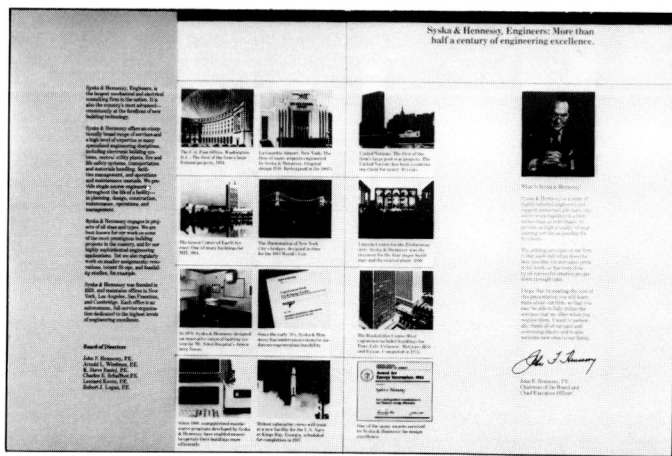

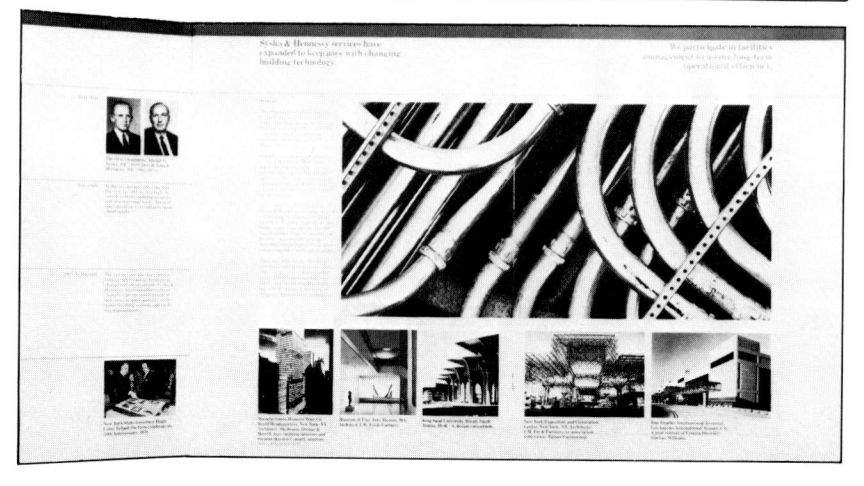

Company Brochure

This bright-yellow brochure shows only the firm's name, in bold black letters. Inside, the front cover continues into a flap. This flap folds out and becomes part of a three-page spread showing small black-and-white photos of projects. Also, a letter from the chairman of the board and a photo are included. At the back of the brochure, a page listing the firm's services is included opposite a full page of type, which on closer examination is a client list.

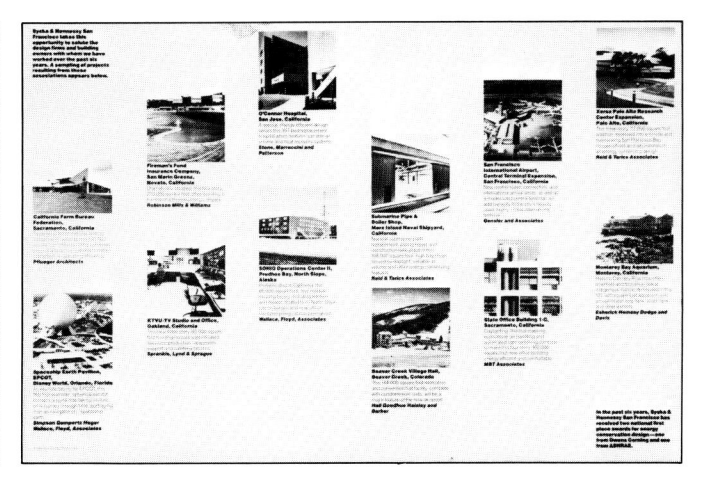

Special Events Piece

To celebrate six years of award-winning work, this firm created a brochure that opens up into a colorful poster. On one side of the piece, a progression of the firm's growth, including major projects completed over the years, is shown through black-and-white cropped photographs with brief descriptive captions. On the other side of the piece is a colorful rendering of a building, taken apart as if in pieces to show the structure of the project.

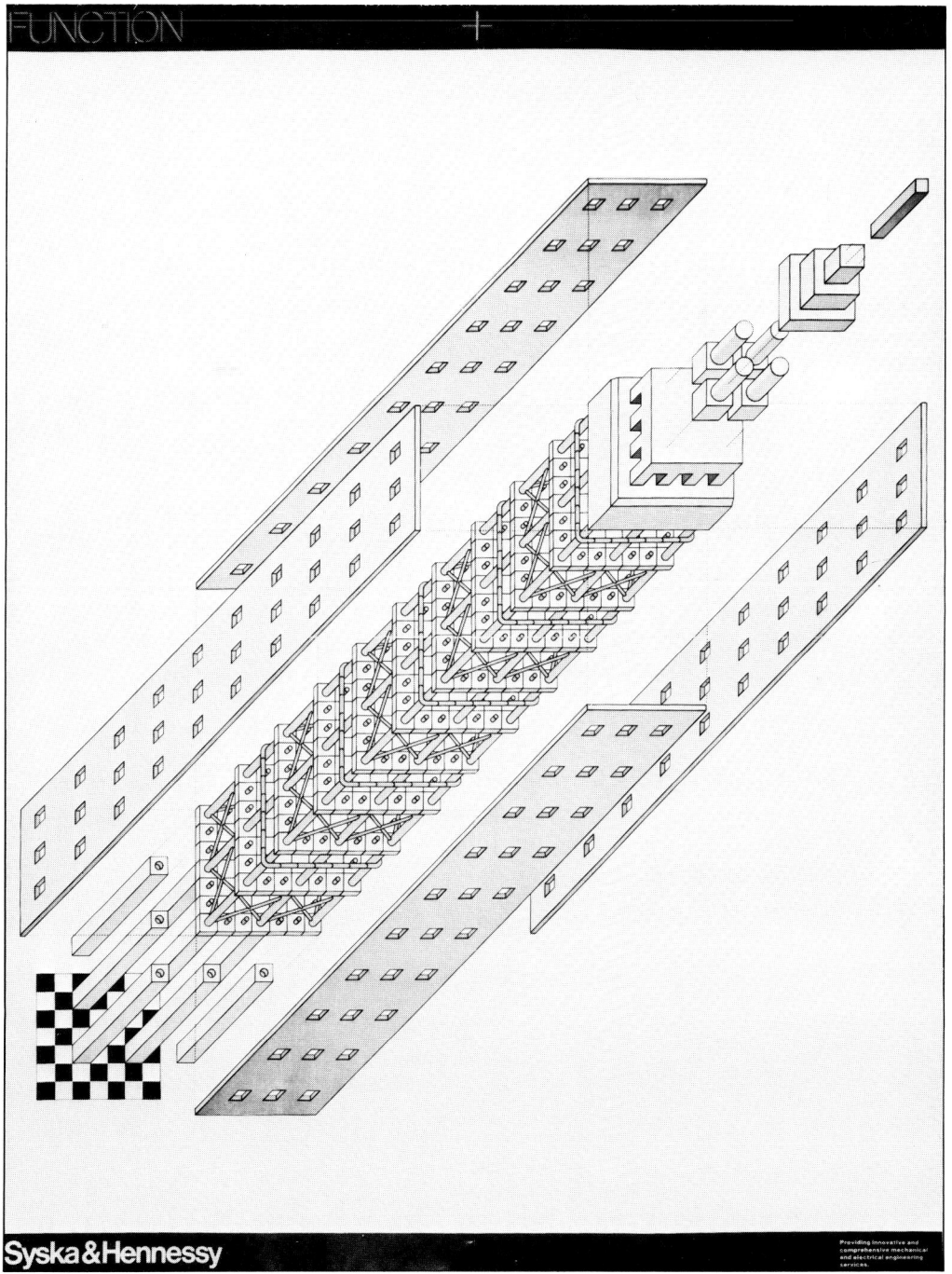

Case Studies: Engineers 175

Baker Engineers

This consulting engineering firm chose a variety of marketing materials for its corporate literature. A new company logo was developed for this material.

Newsletter

This piece features a four-color photo of a recent project on the cover with the firm's name, logo, and a brief list of what is in the issue. The 16-page piece includes an editorial and black-and-white photos of projects and people. The subject matter ranges from questions and answers to brief articles on projects and people.

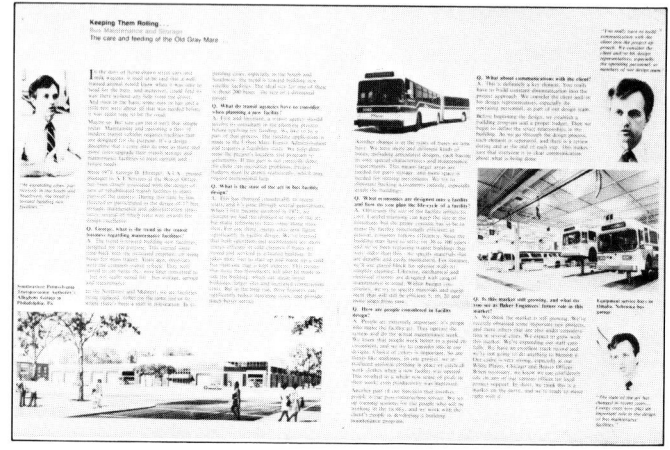

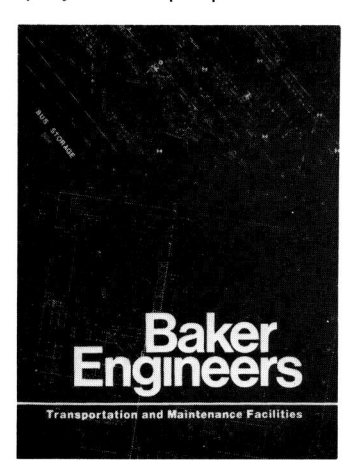

Project Pages

These two-sided pages are black and white, while the blue company logo is the only splash of color. Beneath the photos is a brief description of the project. The reverse side of the page gives specific details of the site. These pages fit into the pocket of a glossy black folder, with white type and orange rule lines.

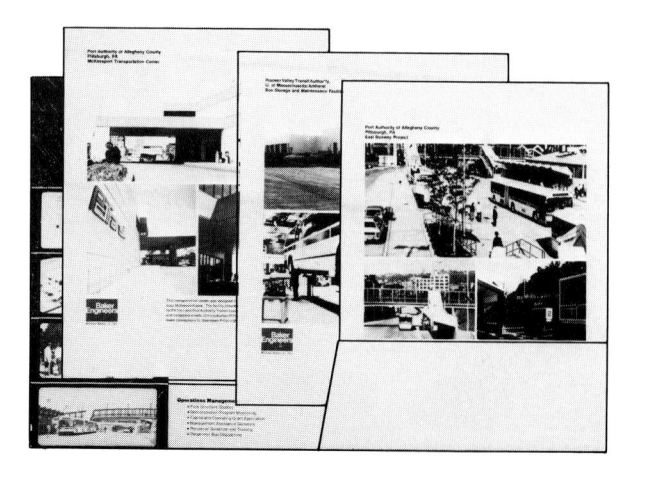

The company uses specialty brochures which detail a particular area. The covers of these pieces have wide stripes in color where all the services are listed. The area of expertise is highlighted in green on each cover.

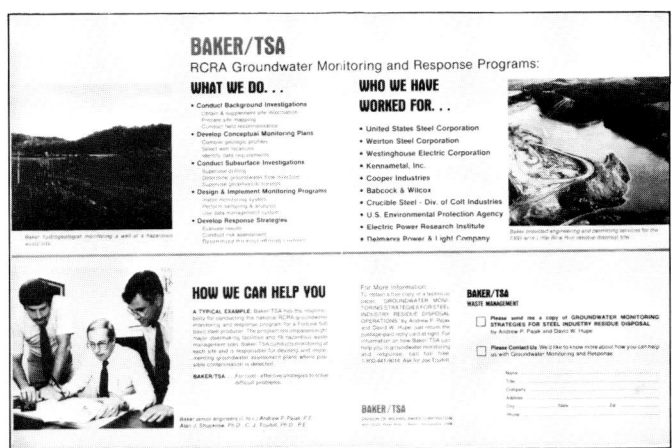

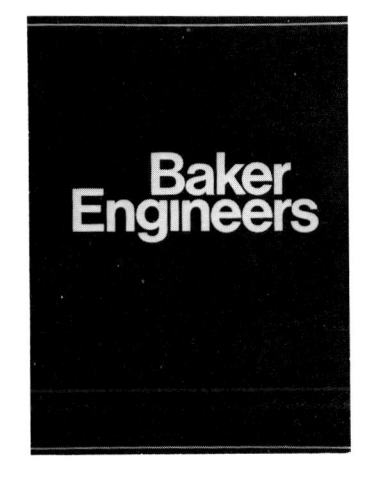

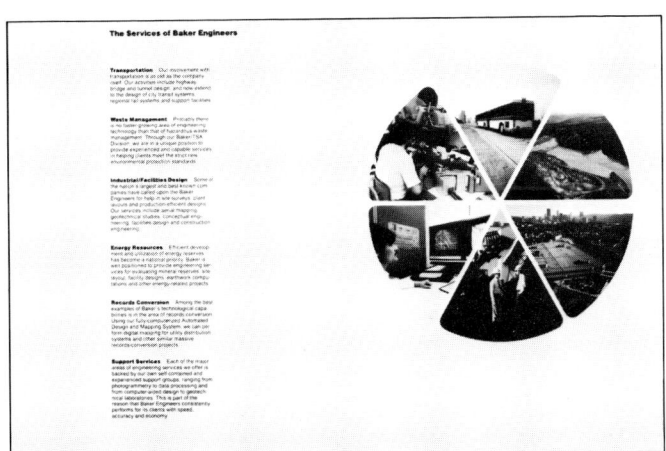

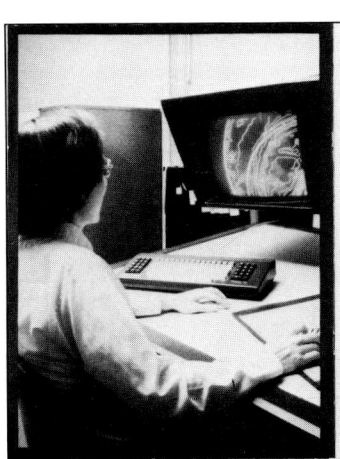

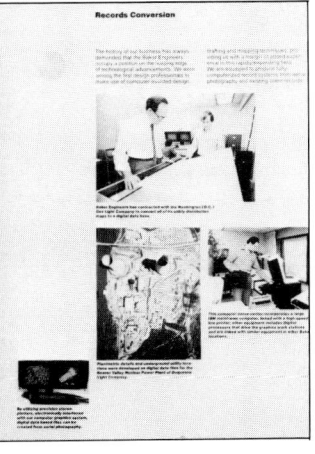

Company Brochure

The glossy black cover has orange rule lines across the top and bottom of the page. The name of the firm is printed in bold white type. The embossed pie-shaped logo design appears here and is repeated throughout the brochure. Inside, the pie appears in light blue. The opening page lists the firm's services. Opposite this page is a pie-shaped graphic enlarged, revealing a different service within each slice. For each section of the book being discussed, an adjoining slice is highlighted in darker blue. Full-page photographs are framed in a black border, and small photographs are used against the white field of the page. On the back of the brochure is a continuation of the orange rule lines, from the front.

Case Studies: Engineers 177

CH2M Hill

This twelve-page quarterly magazine includes articles on the firm's recent projects, and serves as a showcase for the firm's capabilities. Articles are written in a descriptive style which chronicles projects and breakthroughs. Photographs contribute to the visual impact of the magazine. The company logo is repeated throughout the magazine and an insert card is attached to the binding.

Services Brochure

This services brochure outlines the hundreds of services available from this large engineering firm. Several photographs of completed projects and projects in progress are shown and a brief history of the firm is supplied. The back page of this piece gives the locations of the firm's offices, divided by regions.

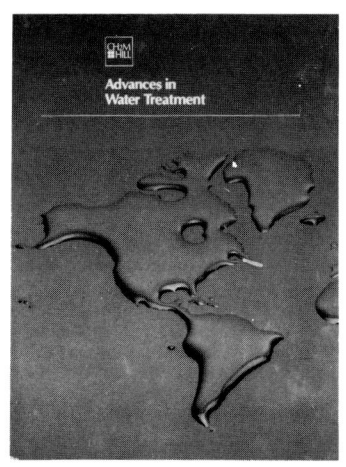

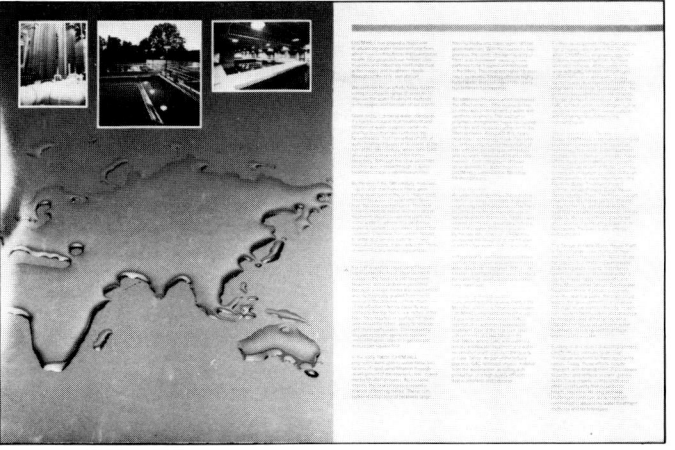

Special Events Piece

This glossy four-color piece features a brochure and poster, which were produced in conjunction with a large historical research project. The cover features water in the shape of a map. This motif is repeated on the interior of the piece, as well as on the poster.

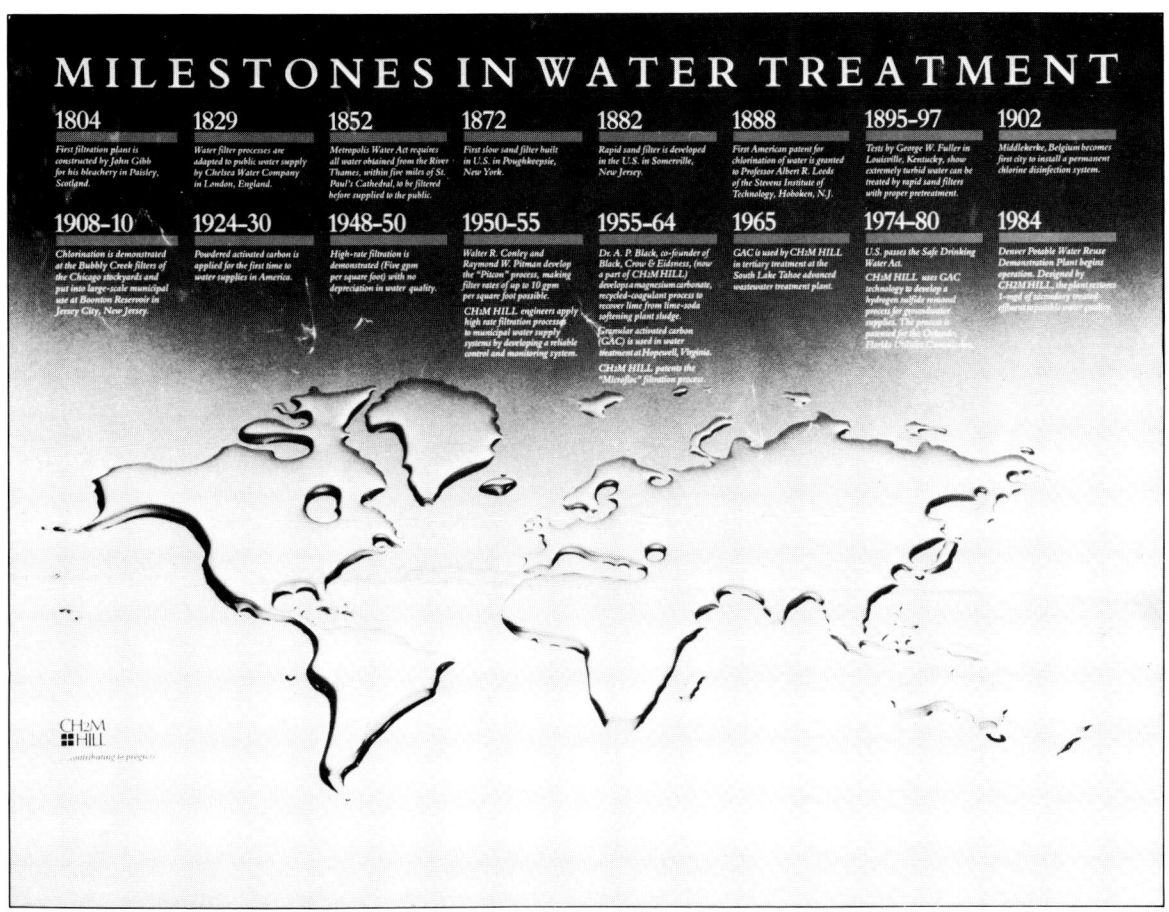

Annual Report

This firm of engineers, planners, economists, and scientists uses an annual report which also serves as a general brochure. The four-color piece addresses itself to past and potential clients, and describes the firm's broad range of services.

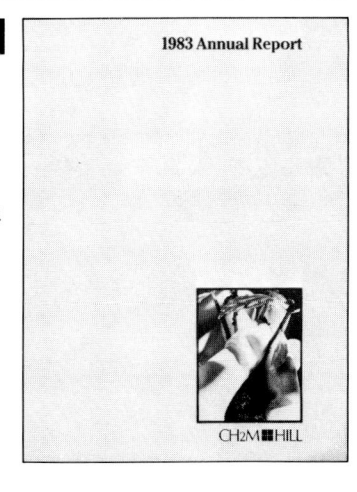

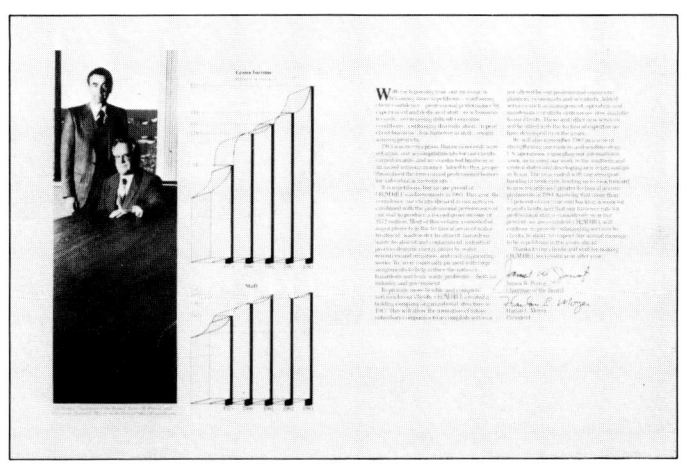

Case Studies: Engineers 179

McElhanney

This six-page newsletter discusses the worldwide work with which the firm is involved. Included are black-and-white photos of the staff and technologically advanced techniques which the firm is using to improve its productivity and expertise. This three-column newsletter also includes renderings and maps of worldwide sites. Editorial comments are also published.

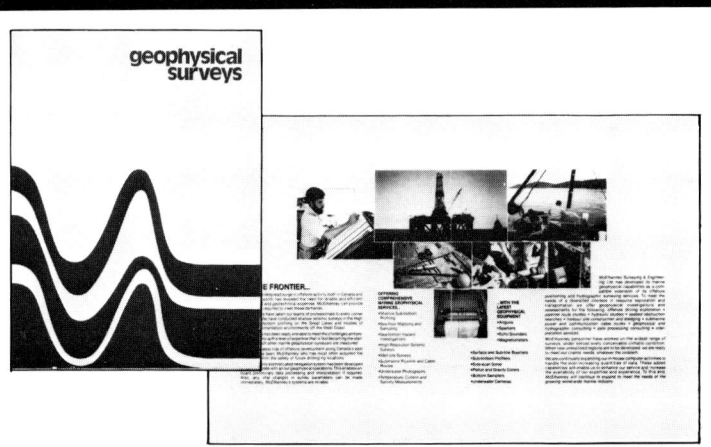

Specialty Brochures

These silver pieces with black type and graphics repeat the silver and black color scheme. The graphics on the front cover symbolize the specialty that the brochure is detailing. Inside these pieces are text about the specialty and color photographs of workers in project situations. The text is brief, highlighting the photos and the firm's expertise.

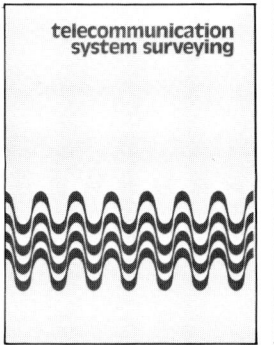

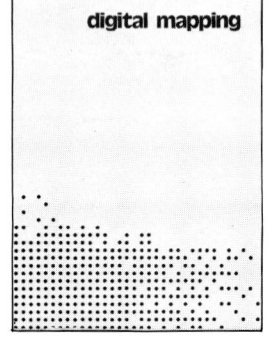

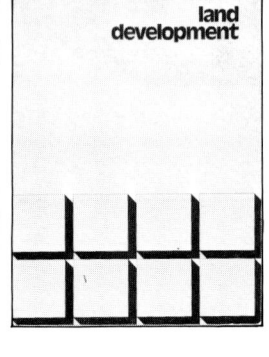

Advertisements

This series of full-page ads combines different elements to showcase its services. Inside the black border a field of silver is shown with computer generated and hand-drawn maps and data dropped out in white. A full-color photograph is also inserted into the silver field. All headline and body copy is printed in the silver area in black.

Company Brochure

For over seventy-five years, this engineering and surveying company has been involved in work around the world. As part of its marketing material, it publishes its main company brochure in several languages.
The cover of this piece is black, with a double front flap cover showing only the name of the firm in silver.

Inside, the brochure includes full-page color photographs of sites. Opposite these pages are full black pages with text in white and silver rule lines. There are also small photographs of details of projects. These visuals, large and small, often show people. These photos are framed in a silver border and silver type.

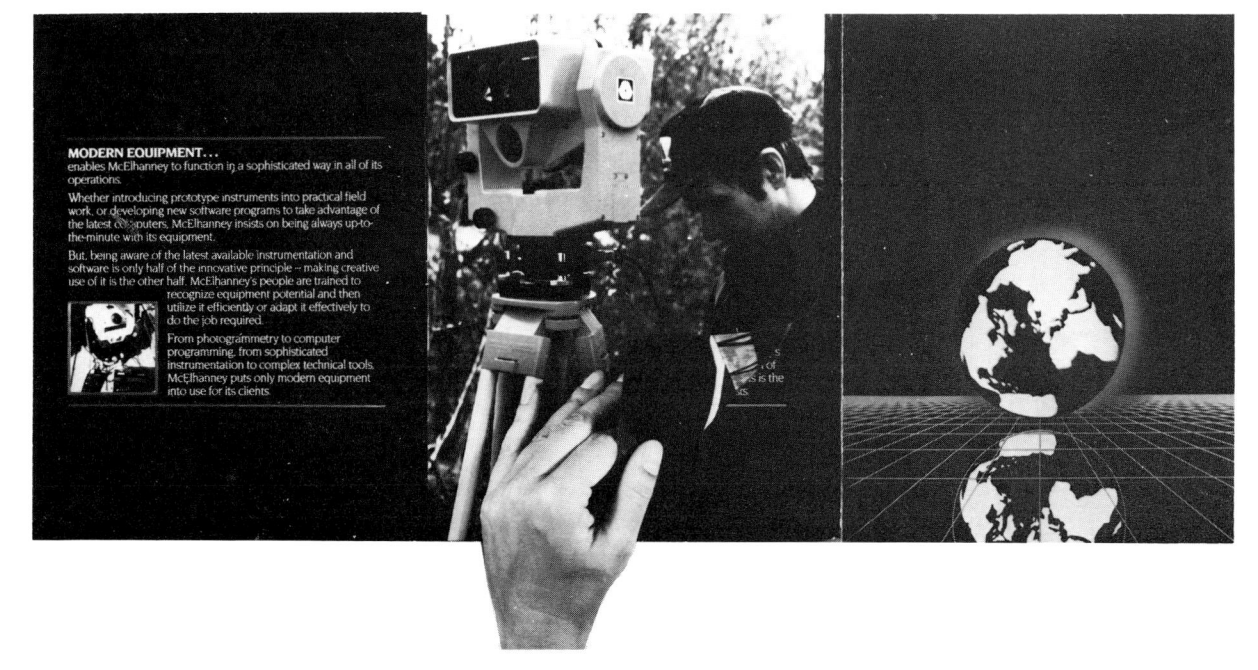

Case Studies: Engineers 181

Harza

Annual Report

This international engineering firm's annual report has color photographs on the front cover featuring its main areas of expertise. Color tabs are used which correlate with different sections of the piece. The report includes project photographs, maps, illustrations, graphs, and charts throughout.

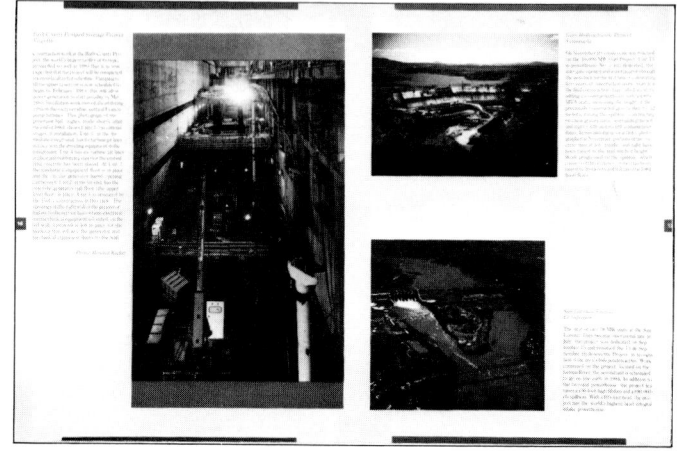

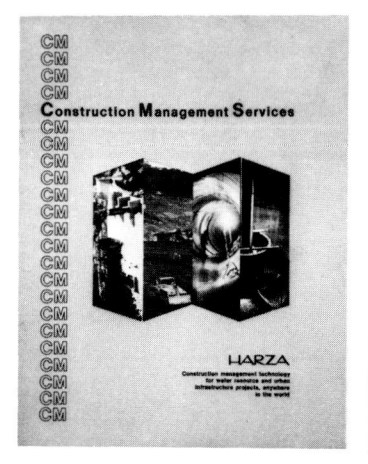

Specialty Brochures

This firm produces thirteen specialty brochures. The specialty brochure on construction management uses photographs on the front cover that illustrate four separate projects in progress. Inside, color photographs and inset photographs are used on almost every page. The text is an integral part of the brochure, explaining the firm's services.

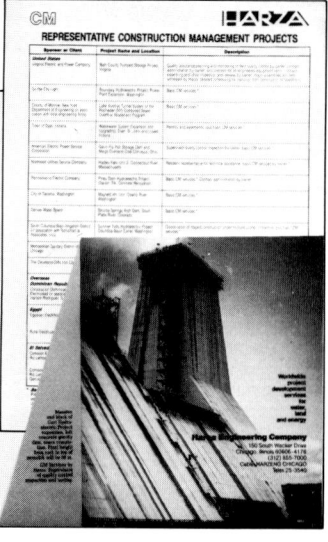

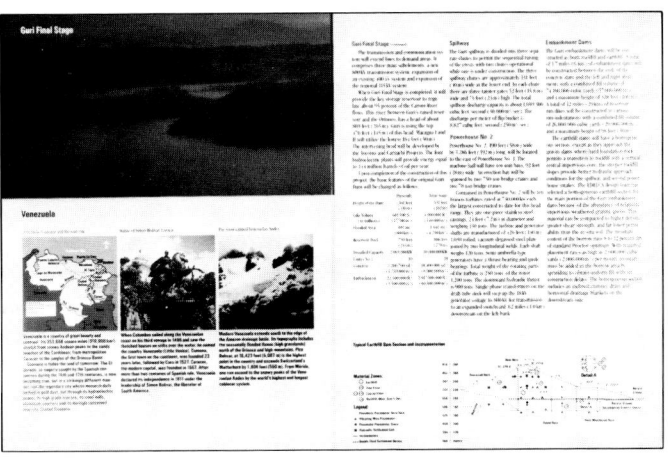

Specialty Brochure

This specialty brochure opens out into a full-sized poster. The outside has a stark black-and-white cover. The type is bold and the copy direct. When the piece is opened, small photos inside show details of the site. The poster opens up fully to reveal a large photograph of the completed project, a map, and a graph showing stages of completion.

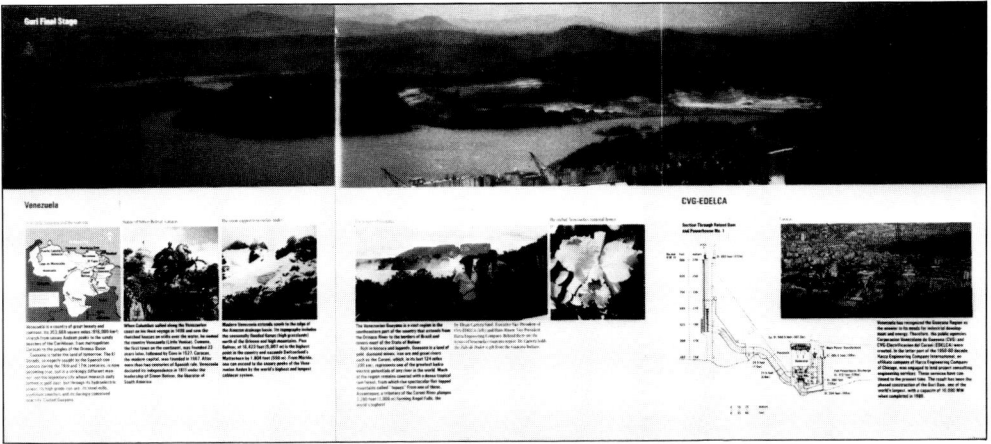

Case Studies: Engineers 183

Sverdrup

This four-page newsletter uses two colors on a glossy cream-colored stock. The newsletter is geared towards highlighting the firm's work and capabilities. Photographs and renderings accompany all articles. This newsletter is used as a self-mailer.

Specialty Brochure

This specialty brochure features a vanishing point grid on its cover with type dropped out in white. The firm's name is printed in gray. This same gray also serves as a background for most of the brochure. The four-column grid is printed directly on the page, and all text and photographs fall within its boundaries.

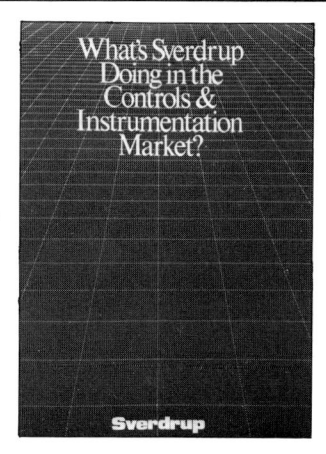

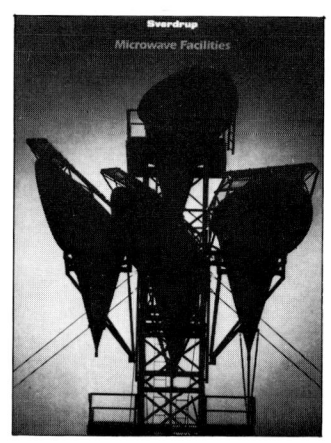

Specialty Brochure

This specialty brochure has a silhouetted photograph of a microwave facility against a violet sky. The firm's name is printed in white and the subject of the brochure is printed in red. The brochure features several pages of people at work and many large full-color photographs of projects.

184

Land Use and Master Planning

1) Queeny Park in St. Louis County, Missouri, featuring 569 acres of recreational property and an award winning children's playground.

2) Airborne and physical training school in Tabuk, Saudi Arabia, featuring 22 buildings to accommodate 900 students and staff.

3) Development of Louisiana Superdome in New Orleans, including master plan for 55 acre site adjacent to central business district.

What's the best way to use the property?

Whether a client's need is to identify the best use of unimproved property or to adapt a particular development to a chosen site, comprehensive land planning capabilities are essential.

Sverdrup's land use and master planning services help a client use property in a manner that is cost-effective, functional, and aesthetically pleasing.

Communities depend on Sverdrup to update or prepare new comprehensive land use plans. These identify appropriate uses based on physical, economic, and social considerations (including historic preservation). Private clients use Sverdrup's highest-and-best-use analyses to maximize a site's utility and financial profitability.

Where a specific development has been selected for a site, Sverdrup can provide a master plan that organizes the project into manageable components, and phases development to take into account current and future needs.

Master plans have been prepared for virtually every type of project: industrial, commercial, and military complexes; industrial parks; new towns; college campuses; recreational parks; airports; ports and harbors; subdivisions; hospitals, and research and development facilities.

Sverdrup's site planning service refines property development. The company locates a facility on a site in the optimum way, taking into account adjacent land uses, existing development patterns, topography, and environmental effects. Sverdrup provides conceptual layouts for all project elements, including utilities, access roads, and landscape features. The company is experienced in preparing rezoning plans, estimating capital costs, and establishing public awareness and acceptance.

Specialty Brochure

This brochure features a brightly colored graphic on its cover. Inside cubes from the cover graphic are repeated with a rule line to start off the text on the page. Full-color, full-page bleed photographs with detail inserts are used throughout this piece. The text discusses the firm's experience.

Project Pages

A glossy folder with a small diagonal grid pattern holds this firm's project pages. When opened, the inside pocket features project photographs printed on the diagonal grid. The corner of the pocket has been cut away so that the title of the project printed on the project page can be viewed without removing the page from the pocket.

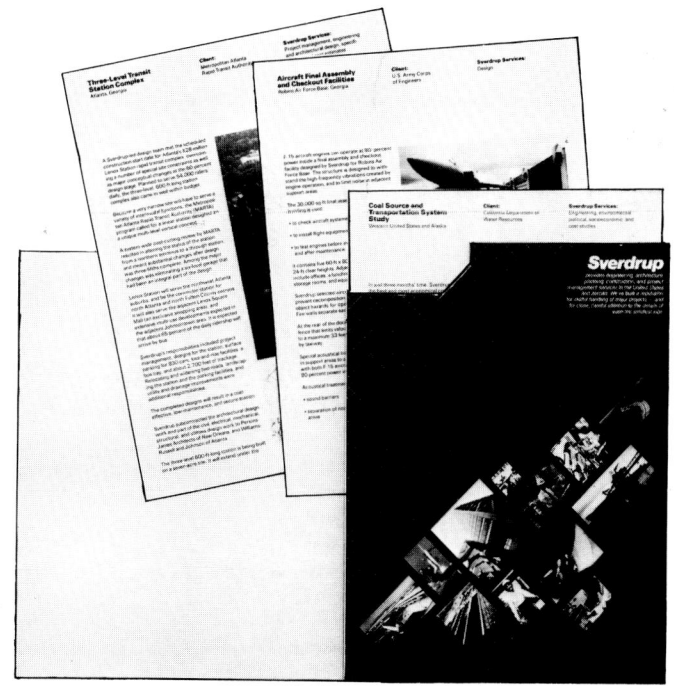

Case Studies: Engineers 185

HNTB

This colorful piece has a traditional magazine format with a four-color full bleed photograph on the cover. The inside cover page lists the firm's partners and associates, and gives information about the cover and the editorial staff. There are usually three to five articles discussing ongoing or future projects in the magazine.

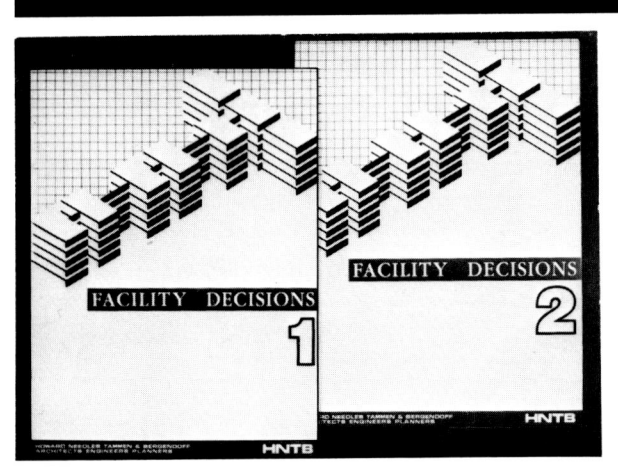

Specialty Quarterly Newsletter

All the covers of these pieces are the same red and black design against a white field. The cover has the name of the specialty on the front. Inside, one main article is highlighted. Photos include black-and-white samples of equipment used by the company; diagrams and charts are also shown.

Discipline Folders

The covers of these pieces are four-color bleeds of projects. The discipline (of the particular folder) and the first initials of the firm appear on the cover. Inside, color photographs and text create a comprehensive explanation of the type of work undertaken by the firm. The back cover includes client information.

186

Specialty Brochures

There are a sizable number of these brochures, which consist of a varying amount of pages and use a perfect binding system. The ½"-wide binding shows the name of the firm and the area of specialty. The front cover has an embossed grid design, which represents the design grid for the inside page layouts. The firm's initials (HNTB) are also embossed on the cover, with the specialty in black letters. Inside, a bright orange page introduces the reader to the firm. The first page is a full bleed color photo of a project. The facing page gives the firm's name, in black against white. The text discusses the areas of expertise, services, and kinds of projects in which the firm is involved. After the initial pages of text, the remainder of the brochure consists primarily of photos of projects, taken from different views (aerial, wide-angle), as well as photos of the models of forthcoming projects. Many of the visuals are in black and white as well as color. At the back of the brochure is a list of clients and awards.

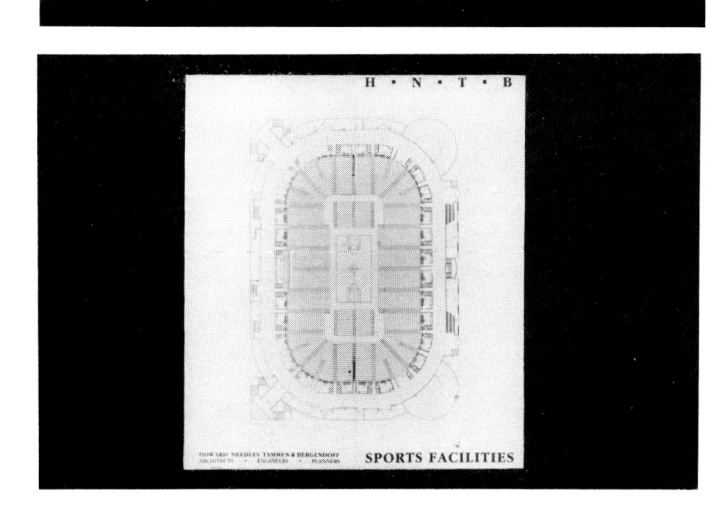

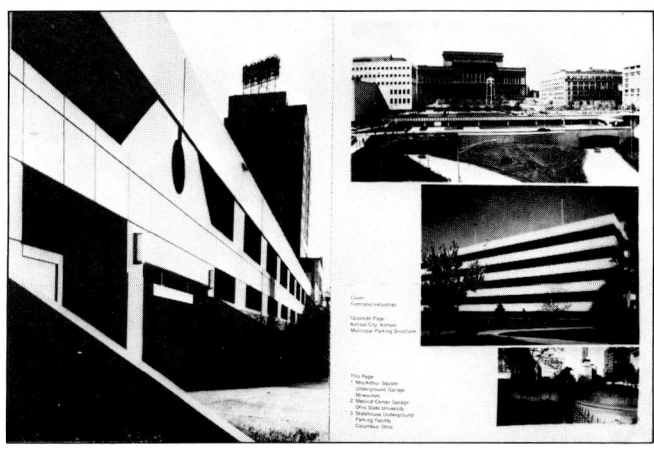

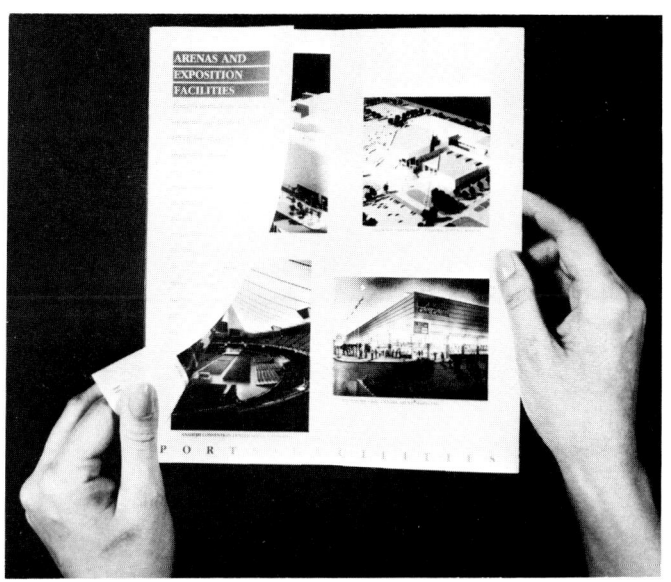

Pocket Brochure

When opened completely, this piece shows a plan view of a sports stadium, in a brown line drawing against a white field. The type on this piece is blue. Inside, color photographs and foldout pages show numerous examples of the firm's work in the particular area. A list of the firm's addresses and some selected projects are given at the back of the piece.

Case Studies: Architects

Parsons Brinkerhoff

This engineering firm has created a magazine that is published twice yearly. A four-color cover is used and the remainder is printed in two colors. A variety of subjects are covered in the publication ranging from staff news to articles on completed projects.

Specialty Folders

These specialty folders use high-contrast photographs printed in one color on the covers. The service featured and the firm's name is dropped out in white type. One large tinted photograph bleeds across the gutter. Several smaller support photographs are shown along with the larger image. The text is kept brief and a capabilities and services list is provided.

Service Brochure

These eight-page brochures are less specific than the specialty folders, dealing with the service as a whole rather than breaking it down into specific project type. The covers of these pieces are printed in one color and the firm's 100-year anniversary emblem is printed in a dull varnish over it. One large full-color photograph which bleeds across the gutter and smaller support photographs are used on many of the spreads. Text and bulleted lists describe the firm's capabilities and services.

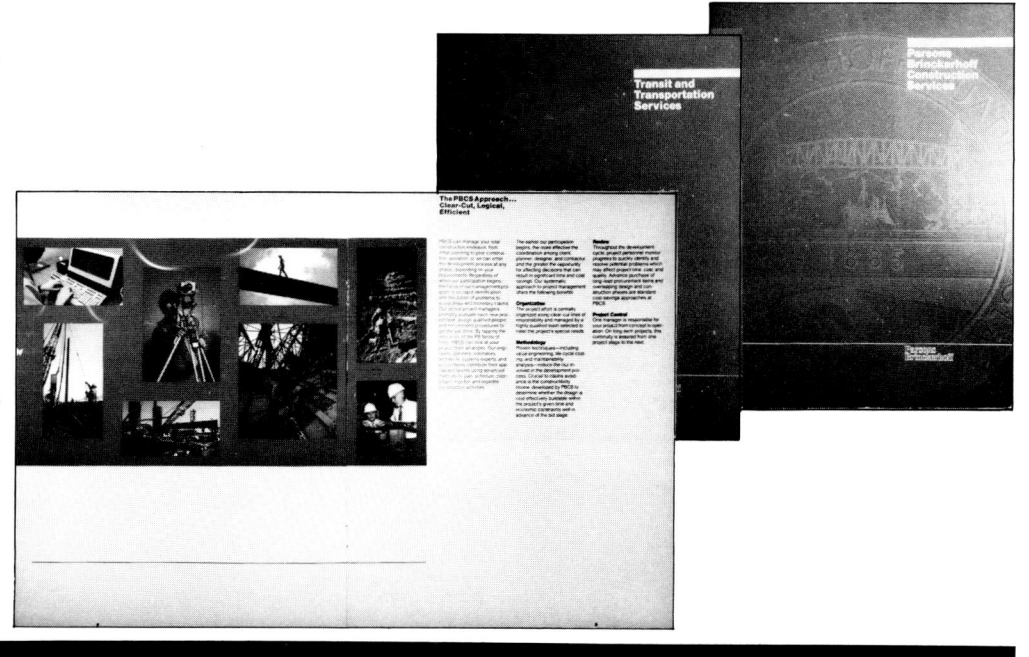

Services Folder

These folders use a varnished emblem for the covers. Inside a series of color and black-and-white photos show people and projects. Bulleted lists that cover all of the firm's services are provided.

Anniversary Calendar

The firm's 100-year anniversary emblem is embossed on the cover of the calendar. The cover and calendar boxes are printed in gold. Photographs of projects and people at work for the firm are used opposite every monthly calendar.

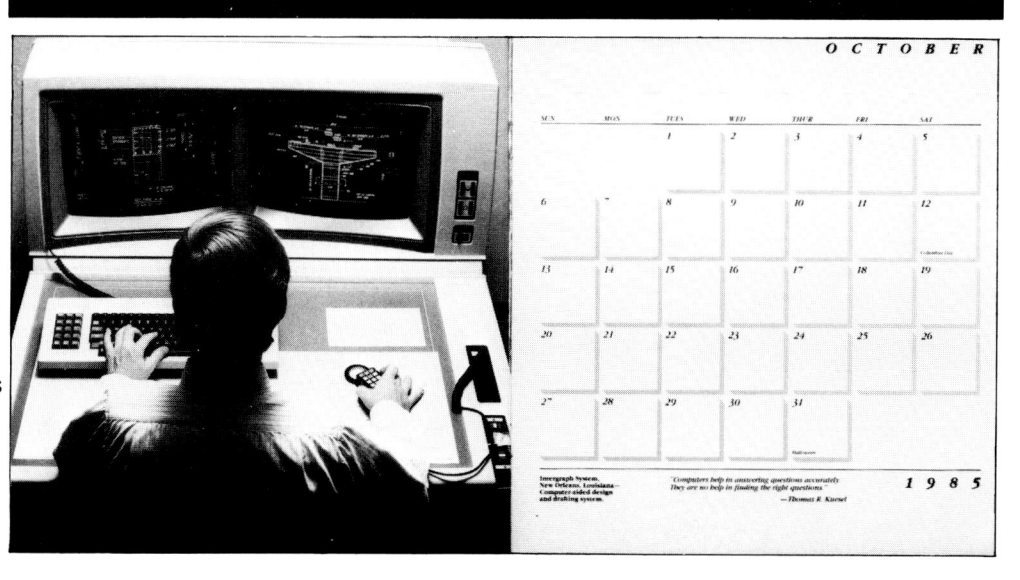

Case Studies: Engineers 189

Opus

These full-color advertisements all employ a sense of humor to get their message across. The headline text and the visuals strengthen the message. Most ads use straightfoward headlines and rely on the photographs to complete the humorous angle. All ads include a photograph of a recently completed project. The body copy describes the advantages of choosing this firm.

Direct Mail

This series of graphically similar, three-page foldout flyers was designed to address issues which most often concern clients. The target audience was fast-growing companies who were faced with the need for additional space. The gray pieces show white graphics with red, blue, yellow, and green on the cover. Inside, the color graphics are repeated. A black-and-white client photo is included with a testimonial quote of satisfaction.

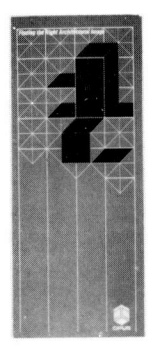

Special Events Piece

This poster was designed as a thirty-year anniversary piece. One side of the poster shows a full-page, four-color bleed photo of a site. Inset detail photos of other sites are set into this photo. The reverse side of the poster details the firm's growth over the past thirty years. The field of the piece is gray with black type and blue numerals. The firm's national locations appear at the end.

Company Brochure

A pale-gray, horizontally striped field with dark blue tabs, white type, and logo (in gold and blue) was designed for this company's brochure cover. Inside, the horizontal stripe pattern is maintained and blue tab rule lines are added. Four-color photographs fit between the lines. White block bullets highlight important points in the text, which discusses the firm and its projects. Photos of the firm's people are used throughout the piece.

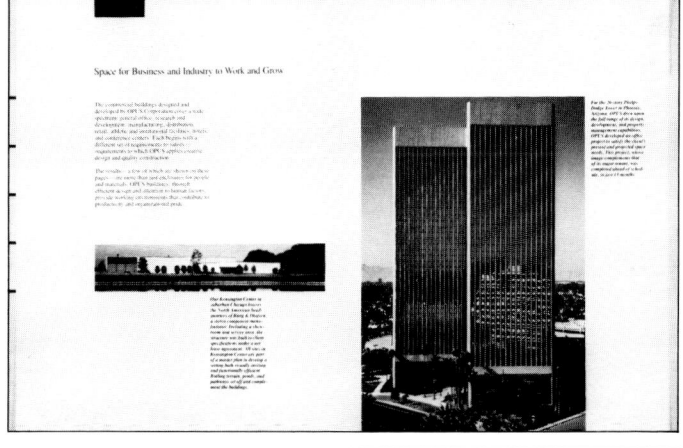

Company Brochures: Construction 191

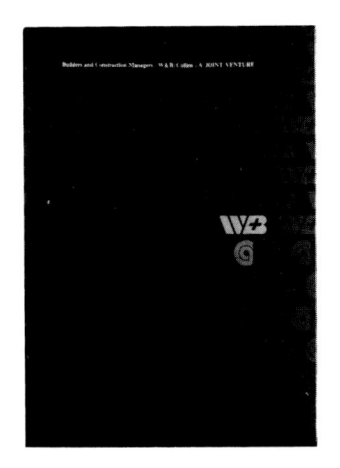

William & Burrows

This special events piece was designed to communicate to potential clients that through a joint venture the firm is able to service geographic areas that it formerly couldn't. The three-fold format allows for the display of several completed joint venture projects as well as a pocket for storing additional information.

Newsletter

This four-page newsletter is produced quarterly and contains information for current and potential clients as well as company members. The articles are kept brief, and the use of accompanying photographs for most of the articles adds extra reader interest.

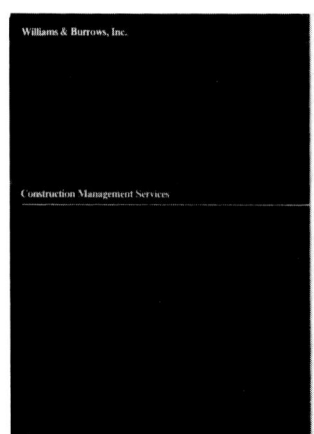

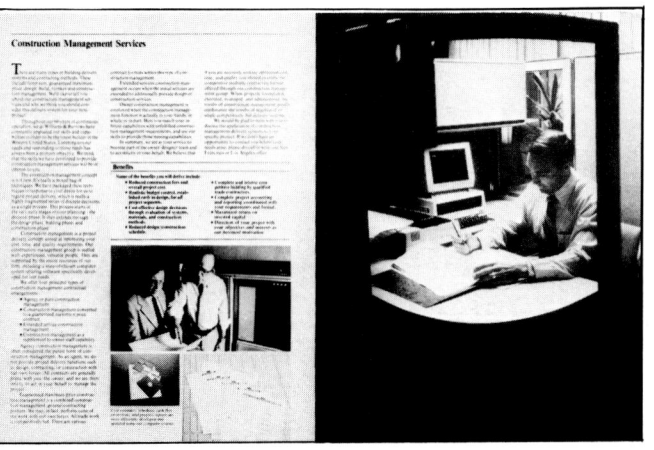

Specialty Brochure

The cover used for this brochure features the firm's logotype shown in a black matt varnish over a glossy black background. The full name of the firm and the specialty area this brochure covers is shown in white dropout type. The brochure discusses services the firm offers. A full-color, full-page bleed photograph with a pocket is featured on the inside back cover.

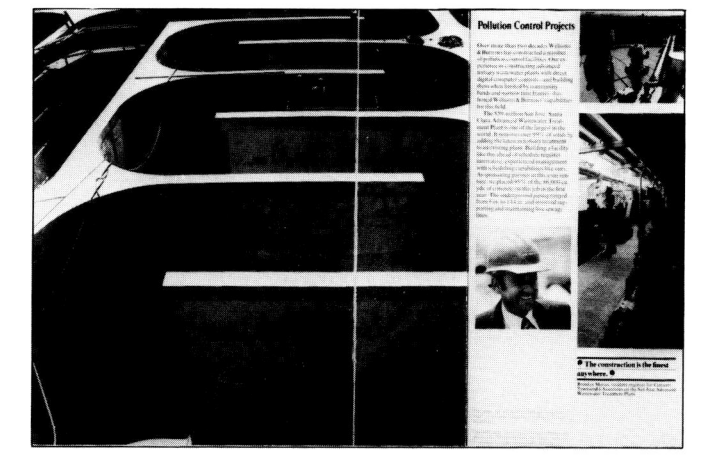

Company Brochure

This firm's main company brochure uses several key elements to express its expertise and experience. Large photographs (which bleed across the gutter) are used for every spread along with smaller photographs of the same project. A very important aspect of this piece is the use of quotes from the firm's clients which stand out on every spread.

Case Studies: Construction 193

Rudolph/Libbe

This construction company has created a brochure system for the firm's marketing purposes. This brochure system and complete client listings have been packaged in a sturdy terra cotta colored box with the firm's name and logo printed in gold lettering on the cover. The liner of the box is a full spread photograph showing several tools of the construction trade. This full-color photograph works well as an invitation to further investigate the contents of the box.

Specialty Brochure

This spiral-bound piece was designed to be a flexible presentation brochure. An embossed graphic of a construction site is used for the cover and all type is shown in the company's colors of terra cotta and black. Dividers are used to separate the different types of projects where company services have been put to use.

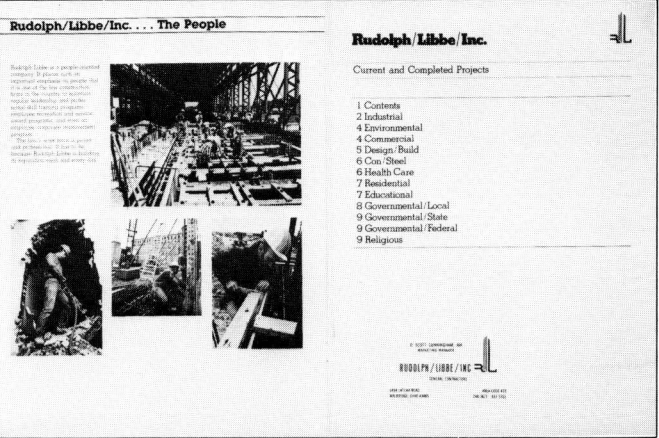

Specialty Brochure

This brochure uses a full bleed color photograph for its cover. This brochure is a smaller version of the main company brochure and is targeted towards the smaller client, so as not to overwhelm. This piece is also used as a handout to presentation boards with multiple members and can be bound into proposals.

Company Brochure

This brochure works as an introduction to the firm for use with major corporate clients, developers and government agencies. The brochure uses a terra cotta cover and gold printing for the firm's name and logo. The textured stock, color and printing make it an exact match to the box it is packaged in. The spiral binding allows for flexibility and updating. Many completed projects are shown in full color. Half-pages with additional text and photographs accompany the standard 8½ X 11 inch pages.

Project Pages

A vertical pocket found in the rear of the brochure provides storage space for a set of project profiles with detailed information for the client to review.

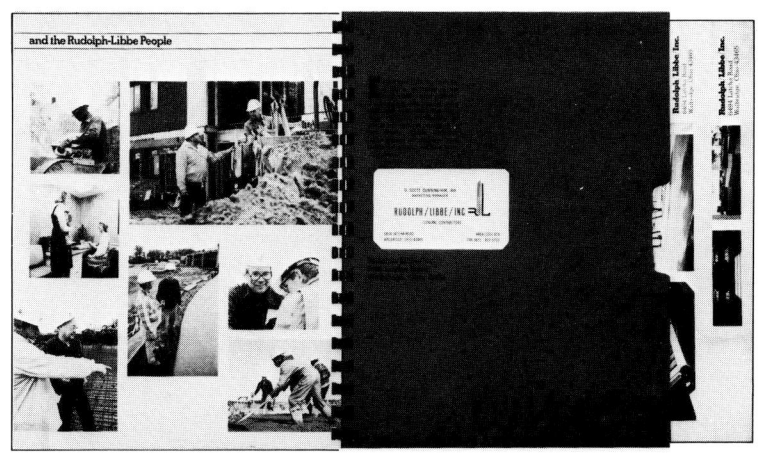

Brochure Systems: Contractors 195

Al Cohen

A black folder of heavy textured stock holds this firms materials. The cover has an embossed logo on it as well as a smaller logo and the firm's name printed in white.

Company Brochure This brochure uses the same cover design as the folder. Inside, project photographs are used along with brief descriptions of the projects featured. Graphic bars are also used to help highlight these photographs.

Project Brochure

This brochure was designed to cover one large project done by the firm. A photograph of the completed project is shown on the cover. Inside, the project is shown on every page in different stages of completion. Quotes and photographs of people involved in the project are seen at the tops of each page.

Newsletter

This firm's quarterly newsletter highlights newly completed projects. Though every newsletter is printed on heavy high-gloss stock, the format changes from month to month varying from a three-panel or four-panel foldout to an eight-page format.

196

Project Pages

The project pages used by this construction firm open from left to right. These full-color pieces show an overall photograph of the project on the cover and detail photographs on the inside spread. A list of the facts is given about the project such as architect, completion time and location. A brief paragraph about the project features and awards is also provided.

Advertisements

These advertisements use a single photograph of a completed project. All the ads in this series use a silver metallic background and brightly colored graphics surrounding the project photograph. The text discusses the firm's capabilities. All the ads include a name and address for further information.

This series of advertisements features a full-page photograph of a completed project and an inserted photograph of the client for that building. A quote from the client is placed next to the inserted photograph and a block of text provides details on the project. The remainder of the text on the bottom of the page consists of statements from the client. These statements are divided into four separate groups: effectiveness, quality, service and the bottom line. The company's name and address are included in the ad, encouraging readers to send for the firm's capabilities brochure.

This series uses full-page photographs showing birds of prey. Each advertisement represents one aspect or strong point of the firm. The descriptive words in the ad relate directly to the strong point of the bird featured. A brief statement follows the name of the firm, and its logo and location are printed at the bottom of the page.

Case Studies: Construction 197

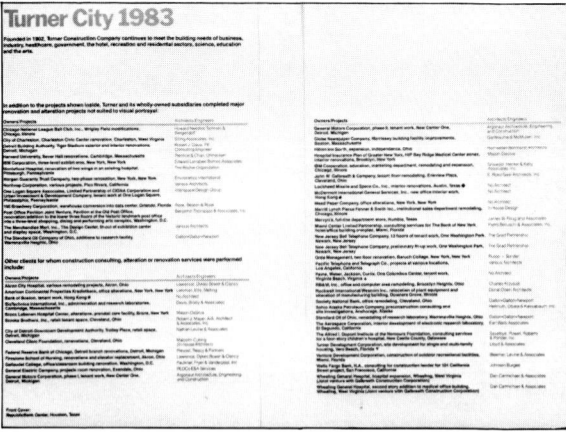

Turner

This worldwide construction firm uses several different pieces for its marketing purposes. This specialty piece entitled "Turner City" has been a company trademark since 1910. The company has created a composite drawing of buildings completed during the preceding year. The buildings shown have been constructed all over the world.

Annual Report

An impressive full-color cover leads the reader into an equally impressive annual report. All financial figures and statements are printed on a coarsely-textured tan stock, while all other company information such as the different disciplines of the firm and the employees who make it all work, are printed on a smooth-textured white stock.

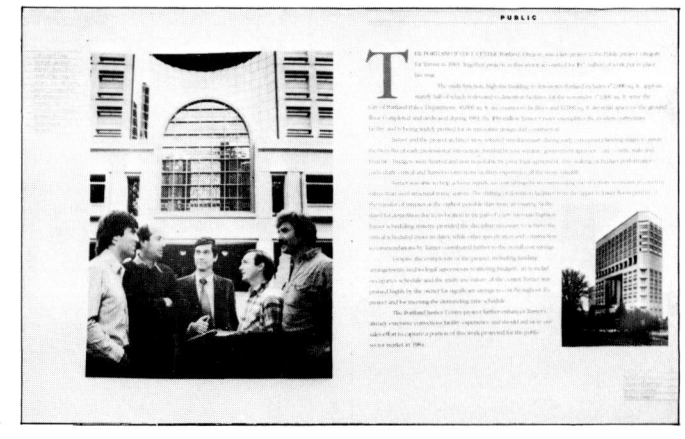

Company Brochure

The cover of this construction company's main brochure shows three men. One is wearing a hard hat with the name of the firm printed in blue against white - the company's corporate colors. The name of the firm is also seen across the top. What comes across is a friendly attitude and one that conveys experience to potential customers.

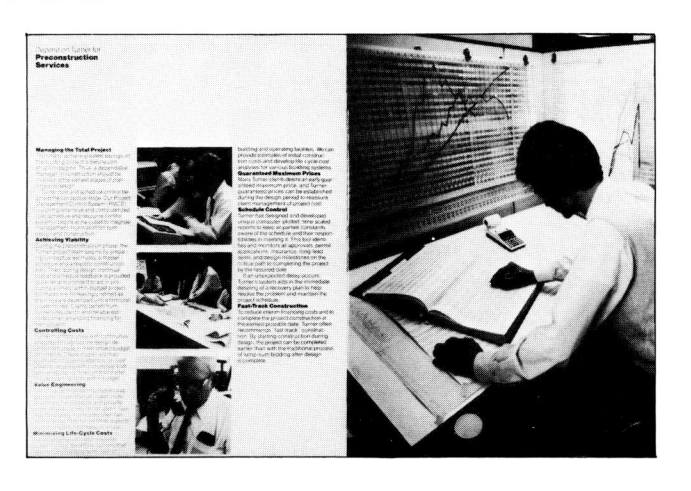

Case Studies: Construction 199

THE ASPEN CLU[B]

Valley Ranch

TIMBER COURT
Professional Office Building

TIMBER COURT
Professional Office Building

INTERTECH
CORPORATE TECHNOLOGY CENTER

SUNLEY HOUSE

East 62nd Street

LEASING BROCHURES: 7

Valley Ranch

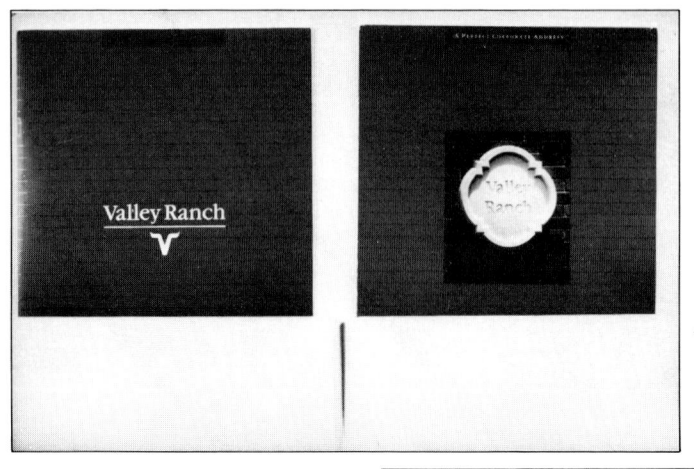

The leasing materials for this residential development were designed with a uniform look. Included in the leasing package are three pieces that fold out into poster size pieces and reveal color aerial views, maps of the area, and maps showing what amenities are available at the location. The same design layout as the main brochure is used throughout the supporting four-color pieces.

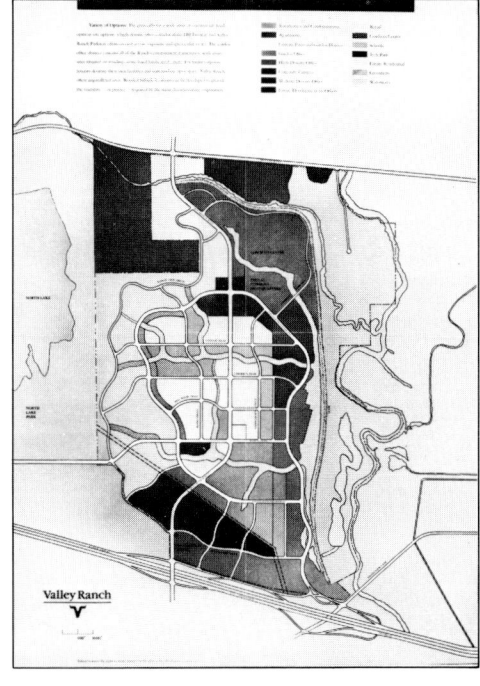

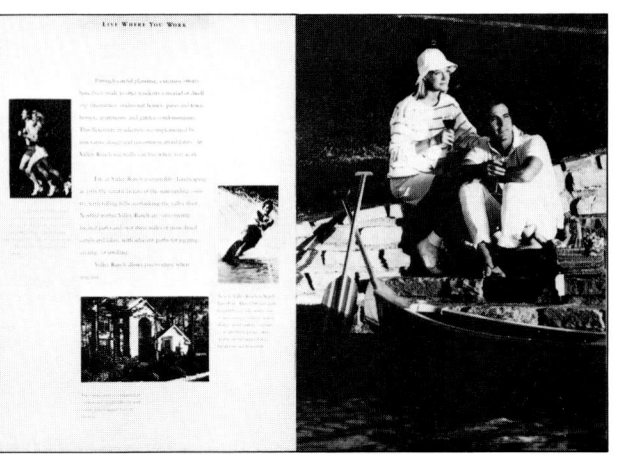

The main leasing brochure has a pale moss green striped pattern on a moss green field. On the inside pages, the tab graphic appears in different colors (beige, blue, peach, green, mauve), designating various aspects of the project. Simple graphs are shown, utilizing the same color as the tabs. Reduced-type captions appear beneath many of the photographs.

Village Center

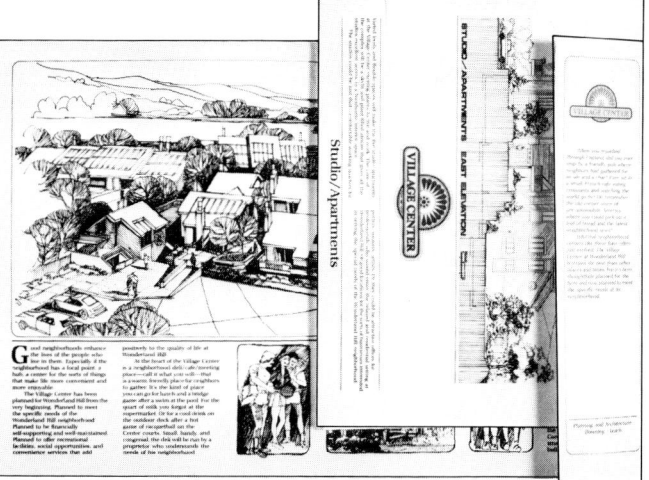

This leasing material is contained in a folder with pockets. The cover of the folder illustrates the project's name and logo used for the signage. The pages stored in the folder are one and two-page detail sheets. These pieces have color renderings of different sections of the property along with the appropriate floor plans. The text is brief, and the prominent logo is used on all pieces.

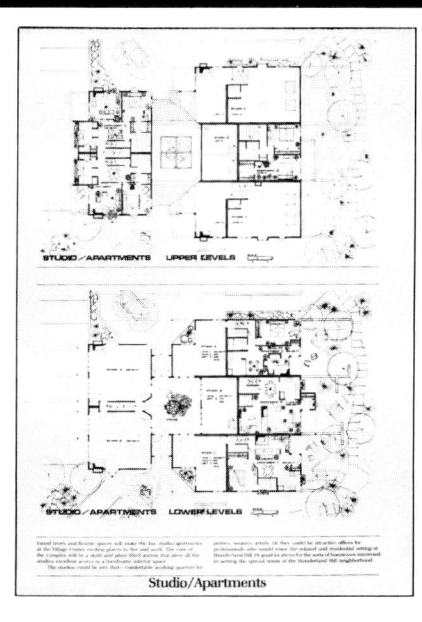

Leasing Brochures 203

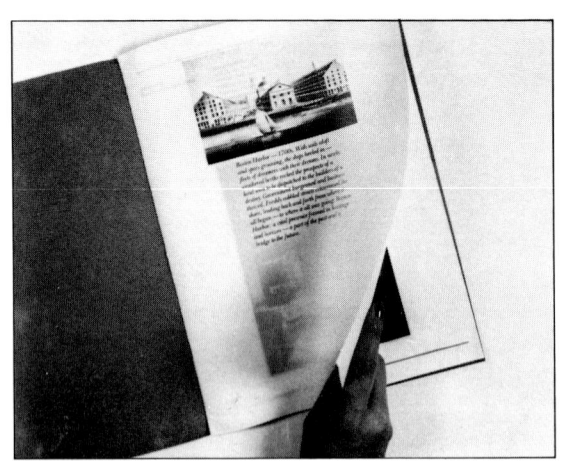

Two Hundred State St.

A gray textured stock was chosen for the cover of this leasing brochure. An embossed graphic of the building is the only visual.
Inside is an acetate overlay with a period etching of the city in which the project will be built. Color renderings and color photographs are used throughout. A reduced image of the embossed building that first appeared on the cover is repeated at the bottom of each page. The text describes the opportunities available in the surrounding area and in the building complex. The back of the brochure has another acetate end paper and a pocket where additional materials can be stored, such as floor plans and other building specifications.

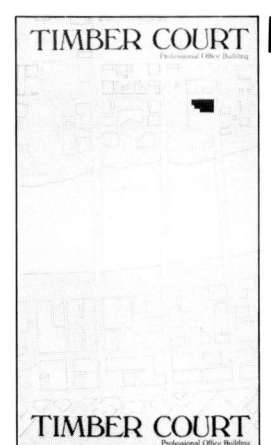

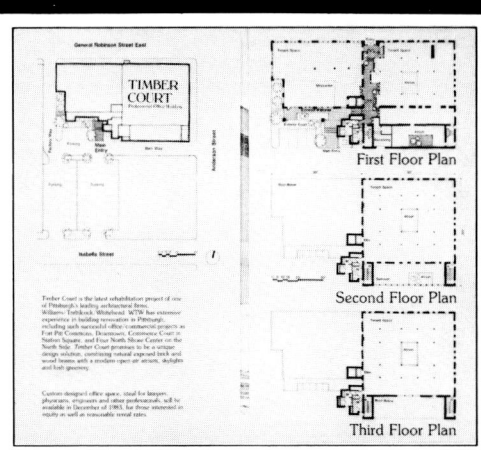

Timber Court

This piece folds out into a two-page spread, then a four-page spread. The cover features a gray-on-white map of the location and area. The location is accented in brown, like the top band of type. The back cover features a map of the area, and provides more information.

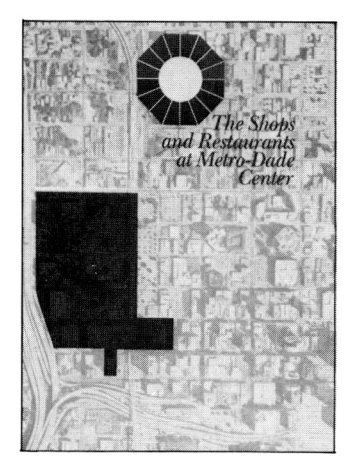

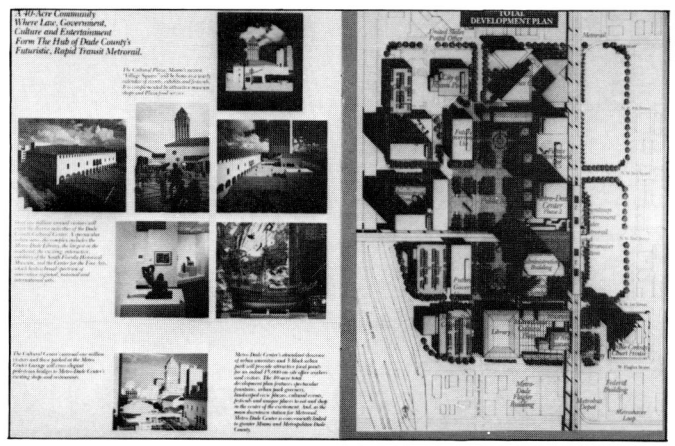

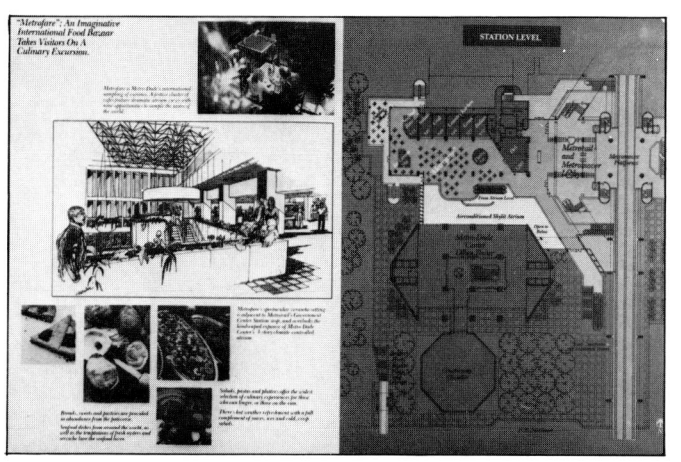

Metro-Dade Center

A black-and-white aerial view of the location of this project creates the field for this cover. A portion of the cover is highlighted in bright green, emphasizing the specific area occupied by this project. The type is a brilliant blue, and the logo is a combination of the blue and green. Inside, color photographs and renderings give the reader a clear picture of the project's location. The text describes the location, the surrounding area, and the opportunities available. A tab line at the top of the page is repeated throughout the brochure. A back pocket is provided, where additional material can be stored.

Intertech

On the cover, a color rendering of a high-tech office building is featured. Inside, a black-and-white reduced print of the cover photograph appears alongside text discussing the building's history and professional capabilities. Renderings and technical drawings are freely used throughout the piece, accentuating the project's position in the high-tech industry.

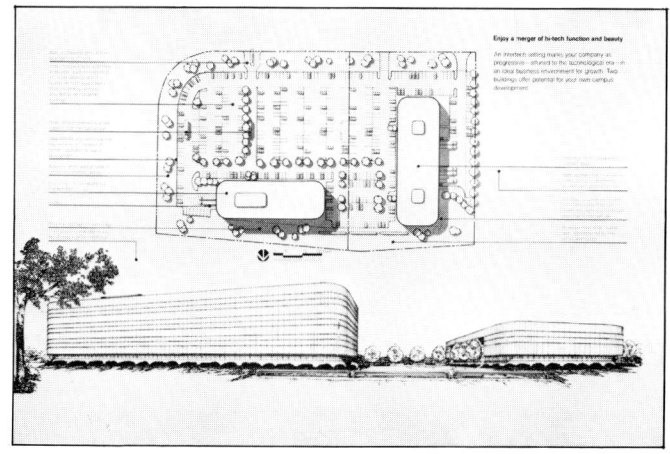

Leasing Brochures 205

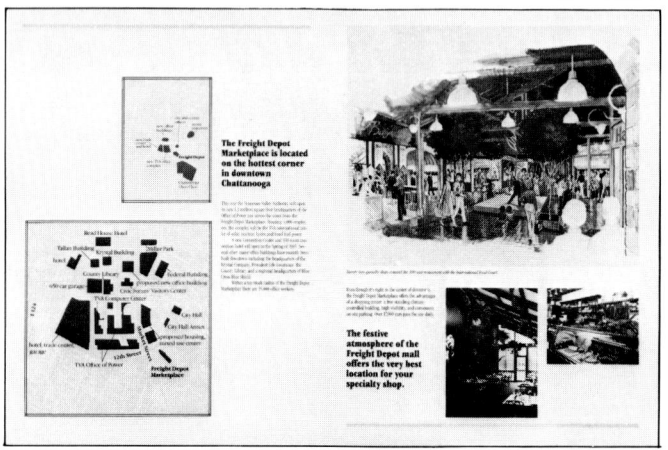

Freight Depot Marketplace

This leasing brochure comes in a glossy white folder, with a framed colored drawing on its cover. The brochure cover also features the same visual. The pale-green tones of the cover are broken up by red and white rule lines. On the inside cover, a description of the project is given. The pages are all framed in a thin black rule line. Red rule lines are also used to divide the visual from the text. Bold black type at the tops of the pages accentuates the otherwise straightforward text. Renderings and maps are also included. At the back of the piece is a pocket containing a blueprint of the project.

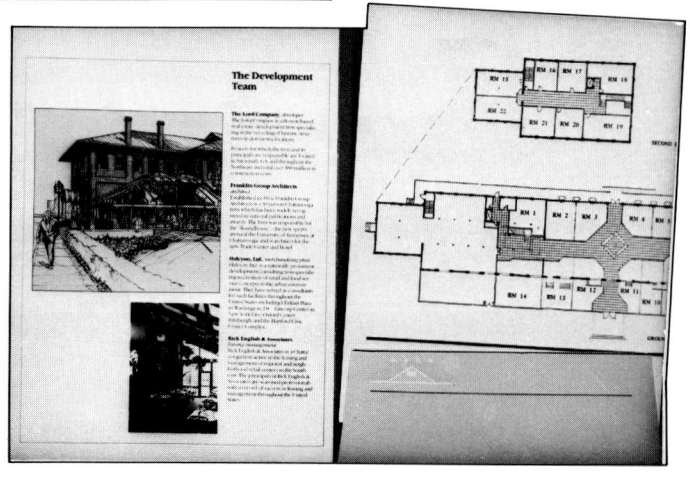

Village Square At Stratton

A forest-green, coarse rag stock was chosen for the cover of this leasing piece. Centered on the cover is a small color rendering. Embossed as well as printed type is used. Inside, full-page color bleed photos and renderings describe the location. A back pocket for additional literature is also provided. The text throughout discusses opportunities and activities that are available at the location. All the paper is heavy quality stock.

The Aspen Club

A beige cover with green rule lines and bold black type distinguishes this leasing brochure. Four-color photographs appear in a grid formation on the cover. Inside, the cover page also shows a color photo. The facing page begins the text, which is printed between green rule lines. Square green bullets are used at the end of the text, and also to accentuate points. The color photos on the cover appear throughout the brochure, and the green rule lines are also repeated.

Leasing Brochures 207

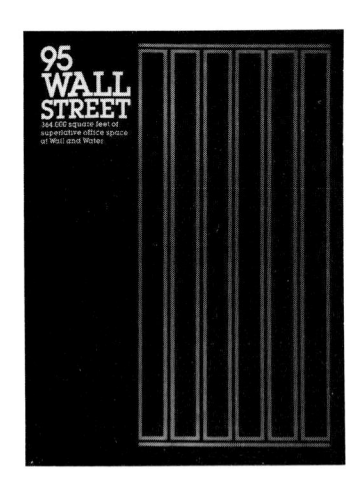

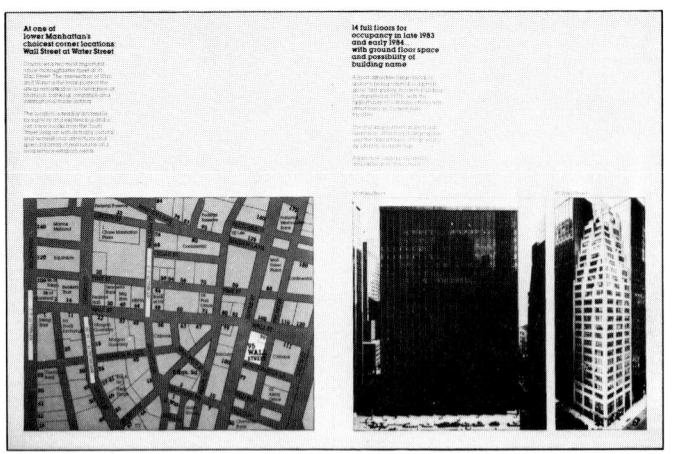

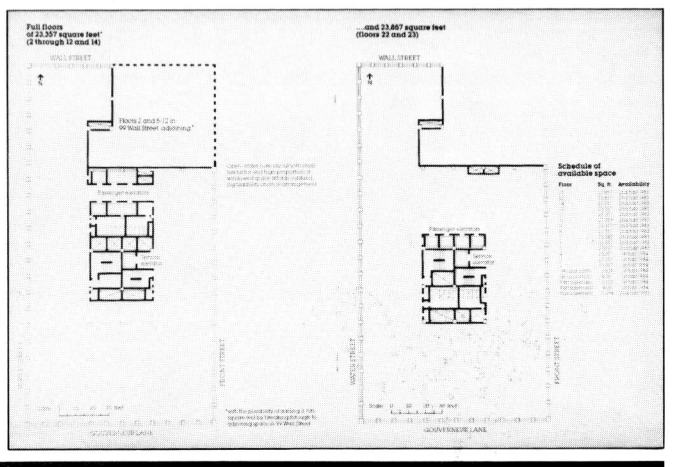

95 Wall Street

This black-and-white leasing brochure shows a frontal elevation of the building being leased. Floor plans and a street map of the surrounding area are included. Brief text describes the project and its important location.

Dallas Centre

This fourteen-page leasing brochure has a dark-blue cover with silver-and-white graphic accents. The back cover shows a rendering of the project. A list of people connected with the project is also included. The four-color full-page bleeds include views of the project, models, renderings, floor plans, interiors, and a small street map.

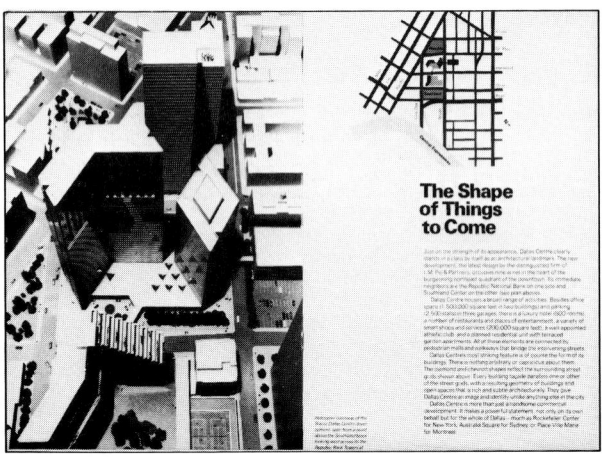

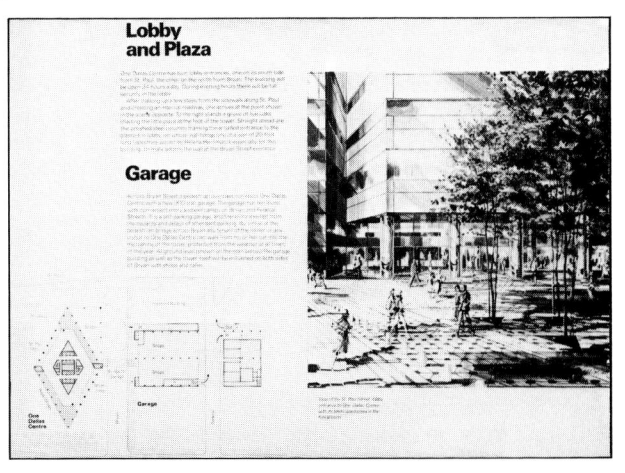

780 Third Avenue

The cover of this three-fold leasing piece features a rendering of the project to be leased and the surrounding buildings and plaza. The rendering is further enhanced by texturing the building's surface. This is done by debossing all of the building's windows. Not only does this enhance the image, it also draws more attention to the building's unique window placement. When opened, large terra-cotta-colored text discusses the projects finer points, completion dates and contact numbers. When opened fully, another rendering is shown and a bulleted list of amenities is given. A flap on the bottom of the center panel holds floor plans. The back cover features a colorful location map and a listing of the architects and engineers involved in this project.

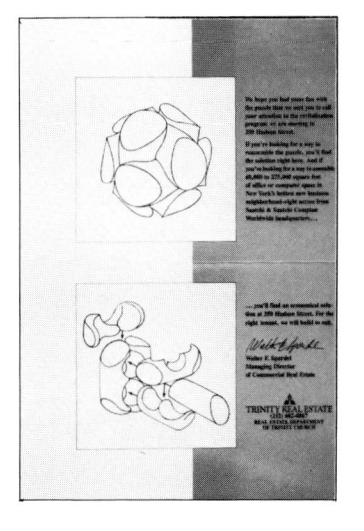

350 Hudson St.

The purpose of these direct-mail leasing pieces was to attract attention, with credibility, to an old industrial building which was undergoing a major reconstruction. Another element was added to this campaign to try to guarantee the serious attention of the leasing brokers. The piece is a high-tech puzzle custom-packaged with a wooden base printed with the 350 Hudson logo. The puzzle and base were then contained in a cloth pouch and boxed in a specially-designed package screened with the same logo. Included in the box was a miniature six-page printed piece relating the assembly of the puzzle to the reconstruction project.

Leasing Brochures 209

Coldwell Banker

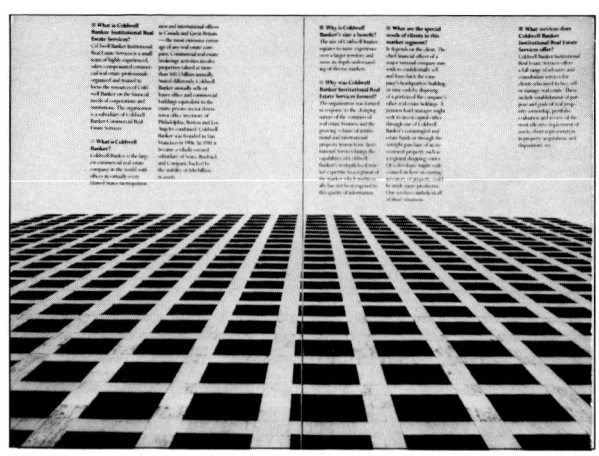

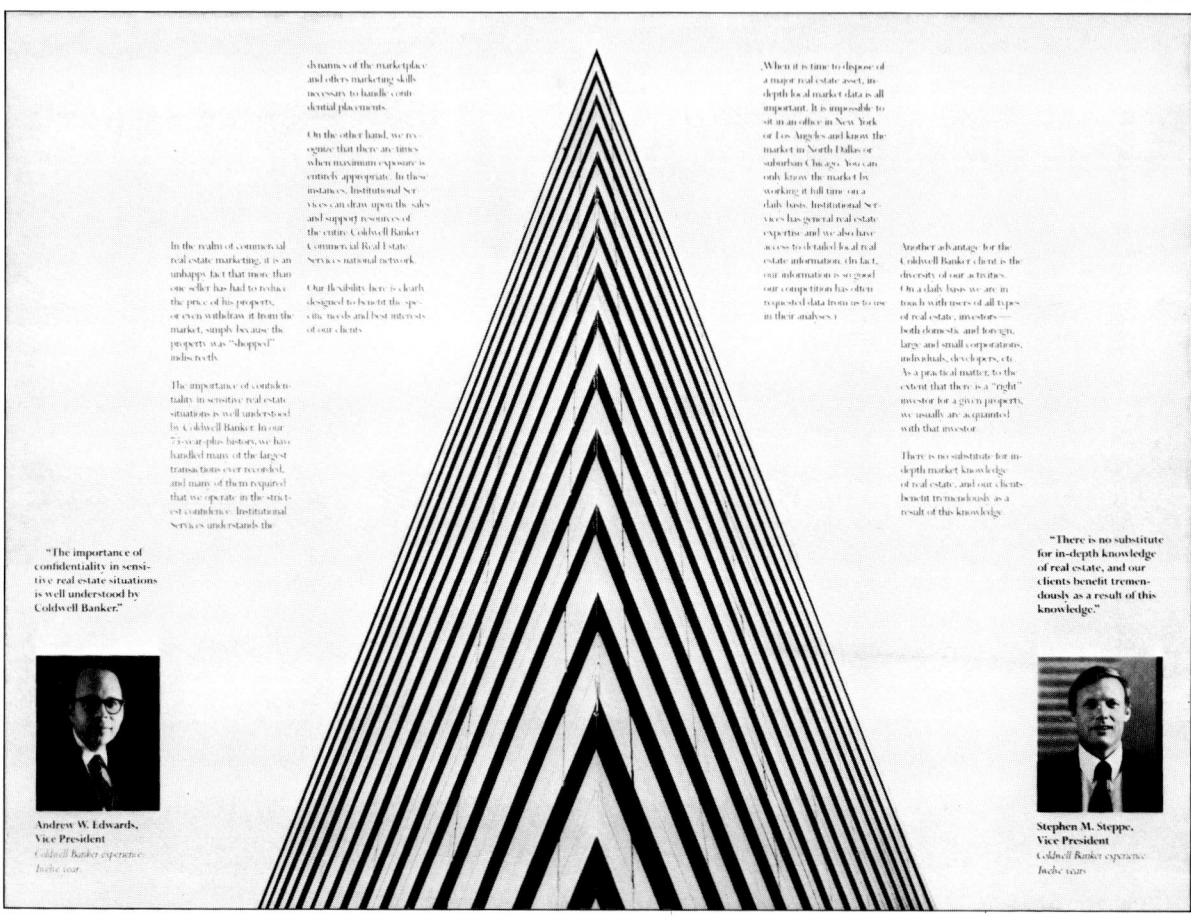

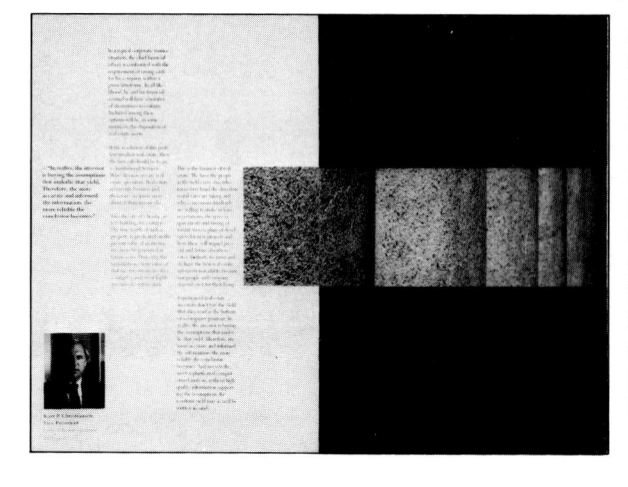

A black mat finish and a full-color photograph make up the cover of this firm's brochure. Glossy black lines are used to continue the building's lines to the edges of the page. The inside front cover is a bright yellow page with one column of text. The text describes why the firm is unique. The facing page contains a letter from the president. On this page, there is also a quote from him, highlighted by a bright yellow square. These squares are seen throughout the brochure, highlighting quotes and questions. Dramatically angled photographs of tall buildings are found on every page, along with photos of key people and bulleted quotes. The text about the firm is not only informative, but is used as a design element in the page layout.

210

The Minskoff Organization

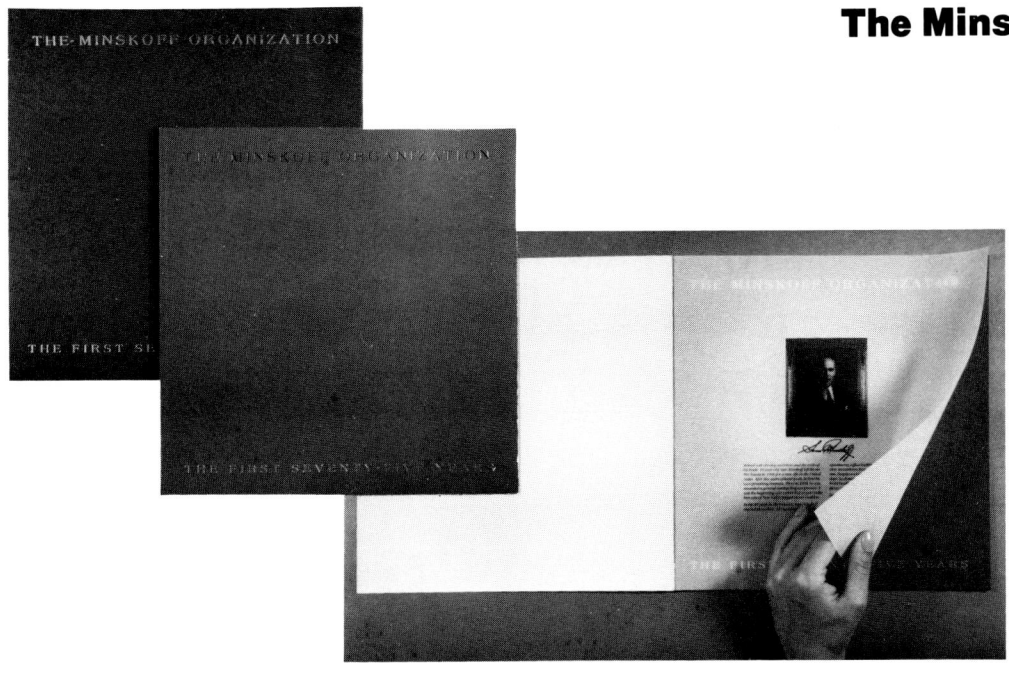

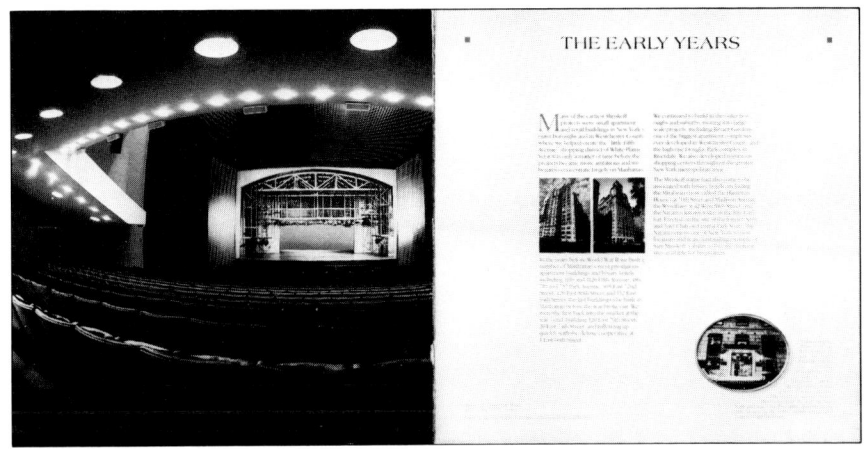

This oversized square brochure comes boxed in a custom-fitted maroon textured sleeve. The sleeve is embossed with the name of the company along the top, and "The first 75 years" on the bottom. This exact pattern and type is also found on the the gray mat brochure cover. Designed as an anniversary piece, this brochure, through words and photographs, takes a trip through the company's first 75 years in business. A frosted vellum overlay starts off this piece - it is imprinted with a portrait of its founder and a brief history of the man and his company. Full page bleeds of instantly recognizable buildings constructed by the firm fill this brochure. Text is opposite, along with smaller photos of projects. On each text page, a bit of nostalgia is seen in the corner, framed in an antique oval frame with accompanying text. Near the end of the brochure, the present Minskoff generation is shown fully involved in this family business. A partial client list is given on the closing piece of vellum.

Leasing Brochures 211

DESIGN CREDITS

14-15
Dewberry & Davis
Fairfax, VA
Design: Corbin Design
Photography: William Mills

16
Washington Associates
Norfolk, VA

17
Rehler Vaughn Beaty &
Koone, Inc.
San Antonio, TX
Design: William Glover Design
Photography: Greg Hursley/
Thom Evans/Cary Whitenton/
William Pilat

24-25
Office Design Associates/
Shepard Martin, Inc.
New York, NY
Design: Burden Associates

26-27
Oliver Design Group
Pittsburgh, PA
Design: Burden Associates

28-29
Alfred Crew Consulting
Engineers, Inc.
Ridgewood, NJ
Design: Burden Associates

30-31
Goldman Sokolow Copeland
New York, NY
Design: Burden Associates

32-33
Rockrose
New York, NY
Design: Burden Associates

34-35
Kirkham Michael And Associates
Omaha, NE
Design: Burden Associates
Photography: Joy Arnold

36-37
Kirkham Michael And Associates
Omaha, NE
Design: Burden Associates
Photography: Joy Arnold

38-39
Kirkham Michael And Associates
Omaha, NE
Design: Burden Associates
Photography: Joy Arnold

48
Index, The Design Group Of
Laventhol & Horwath
Houston, TX
Design: Sharon Tooley Studio
Photography: Frank White

48
Pierce Goodwin Alexander
Houston, TX
Design: Lowell Williams Design

49
Pan Am
New York, NY

49
Charles Caplinger, Planner
New Orleans, LA

50
Ellerbe, Inc.
Bloomington, MN
Design: Ellerbe Graphics Dept.
Photography: Shin Koyama

51
Business Space Design
Seattle, WA
Design: Susan Woodward Wright,
Craig Hanson
Photography: Chuck Kuhn

52
CMC
Alexandria, VA
Design: In-House
Photography: In-House

52
ADD Inc.
Cambridge, MA
Design: Paula H. Briggs
Photography: Warren Jagger/
Steve Rosenthal
Copy: Marilyn LeVine

53
Raymond Hansen Associates
Santa Monica, CA
Design: Don Kingsbury
Cover Art: Carol Hansen Wagner
Photography: Budd Symes

53
Helen M. Moran & Associates
Cleveland Heights, OH
Design: J. Remington & Assoc.

54
Miles Treaster & Associates
Sacramento, CA

54
Brae Construction, Inc.
San Antonio, TX

55
Grahm/Meus Inc.
Newton, MA
Design: Gary Grahm/
Susan Morison
Photography: Paul Ferrino

56
Ewing Cole Cherry Parsky
Philadelphia, PA
Design Alfredo Beron Design

56
Yearwood & Johnson
Nashville, TN
Design: Jerry Murley, Y & J
Photography: Rob Hoffman, The
Phoenix Group
Copy: Joan Link Armour

57
Bower Lewis Thrower/Architects
Philadelphia, PA
Design: Jeanne Derderian/
Sandi Pierantozzi of Derderian
Associates

57
The Ratcliff Architects
Berkeley, CA
Design: Michael Manwaring
Photography: Robert Bauer

58
Henderson Gantz Architects
St. Louis, MO

58
Lea Group
Boston, MA
Design: Harvey Chin
Photography: Steve Schmitt

58
Hammel Green and
Abrahamson, Inc.
Minneapolis, Minnesota
Design: In-House
Photography: Lea Babcock

59
EDAW
San Francisco, CA

60
J.A. Jones Construction Co.
Charlotte, NC
Design: Judy Ferrell
Photography: Various

60
Enterprise Building Corporation
Dunedin, FL
Design: Bud Peck, Specktacular
Publications

61
CE Maguire, Inc.
New Britain, CT
Design: Charles Karno/
Cynthia B. Fontaine
Photography: Frank Giuliani

62
Post Buckley Schuh & Jernigan
Miami, FL
Design: In-House

62
Myklebust Brockman
Associates, Inc.
La Crosse, WI
Design: Elly Hopkins
Photography: Brad Weisbrod/
Lee Randall/Elly Hopkins
Copy: Elly Hopkins/
Sandra Myklebust

63
Lehrer/McGovern, Inc.
New York, NY
Design: In-House

63
CE Maguire, Inc.
New Britain, CT.

64
Gilbert/Commonwealth
Reading, PA

65
SHWC
Dallas, TX
Design: DBG & H Unlimited

66
Honeywell
Minneapolis, MN

66
Transamerica
San Francisco, CA

67
David Paul Helpern
New York, NY
Design: Caplin Communications

67
Gilbert/Commonwealth
Jackson, MI

68
Morris/Aubry Architects
Houston, TX
Design: William Burwell/
Janet C. Goodman

68-69
HLW
New York, NY

72
Burt Hill Kosar
Rittelmann Associates
Butler, PA
Design: Ambit, Incorporated

72
Holt + Fatter
Austin, TX
Design: Joseph J. Holt/
Mervin E. Fatter, Jr.

73
ISD Incorporated
New York, NY
Design: ISD Incorporated/
Bernstein Design Associates

73
Calcara Duffendack Foss
Manlove Inc.
Kansas City, MO
Design: Harmon, True, Pruitt
Advertising/CDFM Architects
Photography: Don Wheeler, Inc.

74-75
Bobrow Thomas
Los Angeles, CA

74
Interspace Incorporated
Denver, CO
Design: Office Staff

75
LZT Associates, Inc.
Peoria, IL
Design: LZT Associates, Inc.

76
Greenboam & Casey
New York, NY

76
Williams Trebilcock Whitehead
Pittsburgh, PA
Design: WTW

77
Psomas & Associates
Santa Monica, CA

77
SHWC, Inc
Dallas, TX
Design: DBH & G Unlimited

77
The Callison Partnership
Seattle, WA

78
Oliver Design Group
Pittsburgh, PA

78
Edwards and Kelcey
Livingston, NJ

79
RTKL Associates, Inc.
Baltimore, MD
Design: RTKL Associates, Inc

79
Einhorn Yaffee Prescott Krouner
Albany, NY
Design: Ruenitz & Co.

80
John Wolcott Associates Inc.
Culver City, CA
Design: Reginald Head/
Bill Wolpert

80
The NBBJ Group
Seattle, WA
Design: The NBBJ Group/
Graphic Design

80
Williams Trebilcock Whitehead
Pittsburgh, PA
Design: WTW

81
Clark Tribble Harris and Li
Architects, P.A.
Charlotte, NC
Design: Mervil M. Paylor/Clark
Tribble Harris and Li Arch., P.A.

81
Preiss Breismeister
Stamford, CT

82
Thompson, Ventulett, Stainback &
Associates Inc.
Atlanta, GA
Design: Reusso + Wagner

83
The NBBJ Group
Seattle, WA
Design: Chris Spivey
Art Direction: John Whitehill-Ward
Photography: Dick Busher/Karl
Bishoff/James W. Brett/Vincent
Vergel de Dios/Michael
Houghton/Mark Cameron
Text: Sharon Batchelor

84
Sikes Jennings Kelly
Houston, TX
Design: Jana Ross & Associates
Photography: Jim Rantala
Writer: Jo Ann Stone

84
Falick Klein Partnership
Houston, TX

84
The Murphy Group
Houston, TX
Design: Gary W. Murphy

85
The Callison Partnership
Seattle, WA
Design: Ellen Ziegler Designers

85
Perez Limited
New Orleans, LA
Design: Wellington Reiter

86
Giffels/Harte, John J. Harte
Associates, Inc.
Atlanta, GA
Design: Kacey Collins
Photography: Giffels/Harte

87
The Grad Partnership
Newark, NJ
Design: Caplin Communications

87
The NBBJ Group
Seattle, WA
Design: Kevin Henderson
Text: Mark Cameron/
Eleanor Walker

89
HDR Systems, Inc.
Omaha, NE

92
Index, The Design Group Of
Laventhol & Horwath
Houston, TX
Design: Sharon Tooley Studio
Photography: Frank White

93
ISD Incorporated
New York, NY
Design: ISD Incorporated/
Bernstein Design Associates

94
Professional Designs Inc
Boston, MA
Design: PDI/Doerr Associates

95
EPR Interior Architecture/
CRSS Inc.
San Francisco, CA

95
SMS Architects
New Canaan, CT
Design: SMS Architects/
Laure Dunne
Photography: Martin Tornally/
Ezra Stoller

95
John Anderson, Inc. and
Interior Environments, Inc.
Chicago, IL
Design: Joel E. Herm
Copy: Linda Legner

96
FMB
Albuquerque, NM

96
DeWild Grant Reckert & Assoc.
Rock Rapids, IA

97
Calcara Duffendack Foss
Manlove Inc.
Kansas City, MO
Design: Harmon, True, Pruitt
Advertising/CDFM Architects
Photography: Don Wheeler, Inc.

98
Shope Reno Warton Associates
Greenwich, CT
Design: Shope Reno Warton
Associates
Photography: Bernard Askienazy

98
Barretta & Associates, Inc.
Boca Raton, FL
Design: Corbin Design/
PL & P Advertising

99
Short and Ford Architects
Princeton, NJ
Design: Stunz-Moore, Inc.
Concept and Copy: Stout Assoc.

100
Allen-Drever-Lechowski
Architects
San Francisco, CA
Design: Assistance from Bruster,
Archer, Ball Advertising
Photography: Gerald Ratto

101
Stephan Lepp Associates
Long Island City, NY
Design: Terry Colbert, Colbert
Design
Copy: Stephan Lepp/Terry Colbert

102
The Architects Collaborative, Inc.
Cambridge, MA
Design: TAC Graphic Design/
Doerr Associates

103
The Stubbins Associates, Inc.
Cambridge, MA
Graphic Design Consultant:
Shephard/Quraeshi

104
Anshen + Allen Architects
San Francisco, CA
Design: Michael Manwaring
Photography: Charly Franklin

105
Peckham Guyton Albers &
Viets, Inc.
St. Louis, MO

105
Anderson DeBartolo Pan, Inc.
Tucson, AZ
Design: Art City
Photography: Richard Payne
Copy: Jo Ann Stone

106
The Ratcliff Architects
Berkeley, CA
Design: Michael Manwaring
Photography: Robert Bauer

107
LZT Associates, Inc.
Peoria, IL
Design: In-House

108
CRS Sirrine
Houston, TX

109
Morris/Aubry Architects
Houston, TX
Design: William Burwell/
Janet C. Goodman

110
Kober Group
Dallas, TX

111
Sugimura + Associates/Architects
Campbell, CA
Design: Peddicord & Assoc.
Photography: Don Jensen/
Robert Martin

112
Flack & Kurtz
New York, NY

113
Gore & Storrie
Toronto, Ontario CANADA
Design: John Adams
Photography: Jim Chambers
Editor: Tom Davey

113
Elson T. Killam Associates, Inc.
Millburn, NJ
Design: American Adgroup/
E. T. Killam Associates, Inc.
Photography: Jeremiah Bean/
Russell Shallieu

114
Culp/Wesner/Culp
Consulting Engineers
Cameron Park, CA
Design: Brian Thomas Associates
Photography: Elise Wienger/
Gabe Palmer/Don Landwehrle/
Don Klumpp/Michael Melford/
Scott Barrow/David White

115
Woodward-Clyde Consultants
San Francisco, CA
Design: GNU Group

116
Emcon Associates
San Jose, CA
Design: Kim Mortyn
Photography: Bill Delzell

116
Wood/Harbinger, Inc.
Kirkland, WA
Design: The Wells Design Group
Photography: Dick Busher/
Richard Bush

117
Schuchart & Associates
Seattle, WA
Design: Philip N. Osborn/
David Peter Greaves
Photography: Bob Thompson/
Louis Bencze/Walter Hodgers/
Davis Holms
Production. Philip N. Osborn

117
Brundage, Baker &
Stauffer, Limited
Columbus, OH

118
Nab Construction Corporation
College Point, NY
Design: Regn/Califano
Photography: Roy Stevens

119
O' Brien-Kreitzberg & Associates
New York, NY
Design: John Brooks
Art/Mechanicals: Etta Jacobs
Creative Supervision and
Copy: Alice Brooks
Photography: Richard L. Lathrop/
Robert Ouan/Peretti & Park/
Roy Engelbrecht

119
Brae Construction, Inc.
San Antonio, TX

120
Tishman
New York, NY

121
Partners Construction
Houston, TX
Design: Partners Construction, Inc.
Photography: Ben Garacci

122-123
Dillingham Construction
Los Angeles, CA
Design: TheMcQuerter Group/
John Benelli/Cognata
Associates, Inc.
Photography: Josh Mitchell

Design Credits 213

124
Griswold, Heckel & Kelly Associates, Inc.
Chicago, IL
Design: Mitchell Cohen

125
Trisha Wilson and Associates
Dallas, TX

125
Hatfield Halcomb Architects
Dallas, TX
Concept/Design: Ender Assoc
Photography: Dan Stroud

126-127
Wittenberg, Delony & Davidson, Inc.
Little Rock, AR
Design: Bruce Wesson/
The Art Department
Photography: Greg Hursley
Text: Susan W. Barr

128-129
Clark Tribble Harris and Li Architects, P. A.
Charlotte, NC
Design: Merviln Paylor/Clark Tribble Harris and Li Architects

130
The Hillier Group
Princeton, NJ

131
Williams Trebilcock Whitehead
Pittsburgh, PA
Design: Barton, Denmarch, Esteban, Inc.
Photography: Dennis Marsico

132
Havens and Emerson, Inc.
Cleveland, OH
Design: Cord & Welentz
Photography: Mort Tucker

133
Burgess & Niple, Limited
Columbus, OH
Design: Lord Communications, Inc
Burgess & Niple, Limited

134
Elson T. Killam Associates, Inc.
Millburn, NJ
Design: American Adgroup/
Elson T. Killam Associates, Inc.
Photography: Jeremiah Bean/
Russell Shallieu

135
Van Dell and Associates, Inc.
Irvine, CA

136
The Earth Technology Corporation
Long Beach, CA
Design: Tony Yazzetta/
Designwise, Inc.
Photography: Herd Kravitz

137
Rock Mountain Geotechnical, Inc
Colorado Springs, CO
Design: Hiesloy Design
Photography: Kenneth Myers

138-139
M.J. Brock & Sons, Inc.
Los Angeles, CA

140-141
Cini-Grissom Associates, Inc.
Potomac, MD

142
Stratton & Company, Inc.
Atlanta, GA

143
Walker Parking Consultants
Kalamazoo, MI
Design/Photography:
Walker Parking Consultants
Graphic Consultant: The Design Center

144-145
Blount International, Ltd.
Montgomery, AL
Design: Marsha Urban/Roy Trimble
Photography: Bud Hunter/Ron Scott/David Dobbs/Photo Options/Four by Five, Inc.

148-149
Vitetta Group
Philadelphia, PA

150-151
Catalyst Incorporated Architecture
Orlando, FL
Design: Peter Shurtz
Production: Peter Shurtz

152-153
Zimmer Gunsul Frasca Partnership
Portland, OR
Design: Design Council, Inc.
Photography: Tim Hursley/
Ed Hershlierger/Various

154
The Callison Partnership
Seattle, WA

155
Hansen Lind Meyer
Iowa City, Iowa
Design: Jack Weiss & Associates

156-157
Perkins & Will
Chicago, IL
Design: P&W Graphics Department

158-159
Gresham, Smith & Partners
Nashville, TN
Design: G S & P Graphics

160-161
Gensler and Associates/Architects
San Francisco, CA
Design: Gensler Graphics Group

162-163
Thompson, Ventulett, Stainback & Associates, Inc.
Atlanta, GA
Design: Reusso + Wagner

164-165
Hellmuth, Obata & Kassabaum
St Louis, MO
Design: Charles P. Reay/Scott G. Hueting/Walter Baetjer
Photography: George Silk/William B. Mathis/Others
Project Management: Anne Henning

166-167
Swanke Hayden Connell Architects
New York, NY

168-169
RTKL Associates, Inc.
Baltimore, MD
Design: RTKL Associates, Inc.

170-171
Brown and Caldwell
Walnut Creek, CA
Design: Robert Conover
Cover Design: Ron Toryfter
Photography: Suzanne and Dave Colwell

172-173
Donohue, Engineers & Architects
Sheboygan, WI
Design: Jody R. Herr
Copy: Stephanie J. Reith/Grace M. Rindfleisch

174-175
Syska & Hennessy, Inc. Engineers
San Francisco, CA
Design: Michael Vanderbyl

176-177
Michael Baker Corp.
Beaver, PA
Design: Robert A. Smith
Copy: Georgia G. Zeedick/Robert A. Smith

178-179
CH2M Hill
Denver, CO
Design: David Hampton/
Tandem Design
Photography: Greg Gorfkle/Staff

180-181
The McElhanney Group Ltd.
Vancouver, B.C., CANADA

182-183
Harza Engineering Co.
Chicago, IL
Design: Dawn McIntosh

184-185
Sverdrup Corporation
St. Louis, MO
Design/Photography: Marketing Support Services

186-187
Howard Needles Tammen & Bergendoff
Kansas City, MO

188-189
Parsons Brinkerhoff
New York, NY
Design: PB Communications

190-191
OPUS Corporation
Minneapolis, MN
Design: Design Center Inc./
Bonnie Richter/Larsen Design Inc./Richard Fischel
Photography: Shin Koyama/Paul Schaumberg/Pat Siegrist

192-193
Williams + Burrows, Inc.
Belmont, CA

194-195
Rudolph/Libbe/Inc.
Walbridge, OH

196-197
Al Cohen Construction Co.
Denver, CO
Design: Schenkin/Assoc., Genesis
Photography: Jay Simon/
Ed La Casse

198-199
The Turner Corporation
New York, NY

202
Valley Ranch
Dallas, TX
Design: Halcyon LTD

203
Village Center
Boulder, CO

204
Timber Court
Pittsburgh PA

204
200 State Street
Boston, MA
Design: Halcyon, LTD.

205
Intertech
St. Louis, MO

205
The Shops And Restaurants Of Metro-Dade Center
Miami, FL
Design: Halcyon, LTD.

206
Freight Depot Marketplace
Chattanooga, TN
Design: Lord Communications

207
The Village Square At Stratton
Design: Halcyon, LTD

207
The Aspen Club
Aspen, CO

208
95 Wall Street
New York, NY

208
Dallas Centre
Dallas, TX

209
780 Third Avenue
New York, NY

209
350 Hudson Street
New York, NY
Design: J.P. Lohman Organization, Inc./Joni Holst/Charles Perry
Illustration: Lee Harris Pomeroy Associates, Architects

210
Coldwell Banker Institutional Real Estate Services
Design/Photography:
The GNU Group

211
Minskoff Organization
New York, NY
Design: Peggy Ford-Fyffe/
J.P. Lohman Organization, Inc.
Photography: Dennis Cox/Bill Miller/Peter Loppacher/Peter B. Kaplan

INDEX

Accordion Folded Mailers, 48-49
ADD Inc., 52
Advertisements, 62-65, 68-69
Al Cohen Construction Co., 196-197
Allen Drever Lechowski Architects 100
Anderson DeBartolo Pan, Inc., 105
Anniversary, 82-83
Anshen + Allen Architects 104

Baretta & Associates, Inc., 98
Blount International, Ltd., 144-145
Bobrow Thomas, 74-75
Bower Lewis Thrower/Architects, 57
Brae Construction, Inc.,54,119
Brochure Evaluation, 42-44
Brown and Caldwell 170-171
Brundage, Baker & Stauffer, Limited, 117
Burgess & Niple, Limited, 133
Burt Hill Kosar Rittelmann Associates, 72
Business Space Design, 51

Calcara Duffendack Foss Manlove, Inc., 73, 97
Calendars 76-77
Caplinger, 49
Case Studies, 146-199
Catalyst Inc. Architecture, 150-151
CE Maguire, Inc.,61.63
Change of Address, 73
Change of Name, 72
CH2M Hill 178-179
Cini-Grissom Associates, Inc., 140-141
Clark Tribble Harris and Li Architects, P.C., 81, 128-129
CMC, 52
Coldwell Banker, 210
Communications Programs: 1-17
Company Brochures, 32-33, 52-53, 90-145
Corporate Brochure, 36-37
Corporate Communications, 2-3
Corporate Identification Manual, 12-13
Corporate Identity, 14-17
Covers, 40-41
CRS Sirrine, 108
Culp/Wesner/Culp, 114

Dallas Centre, 208
David Paul Helpern, 57
Dewberry & Davis, 14-15
Dewild Grant Reckert & Associates, 96
Dillingham Construction 122-123
Donohue, Engineers & Architects, 172-173

EDAW, 59
Edwards and Kelcey, 78
Einhorn Yaffee Prescott Krouner, 79
Ellerbe, Inc., 50
Elson T. Killam Associates, Inc. 113, 134
Emcon Associates, 116
Enterprise Building Corporation, 60
EPR Interior Architecture/CRSS Inc., 95
Ewing Cole Cherry Parsky, 56
Expandable Brochure, 30-31

Falick Klein Partnership, 84
Flack & Kurtz, 112
FMB, 96

Format, 28-29, 44
Freight Depot Marketplace, 206

Gensler and Associates/Architects, 160-161
Giffels/Harte, John J. Harte Assoc., Inc., 86
Gilbert/Commonwealth, 64, 67
Goldman Sokolow Copland, 30-31
Gore & Storrie, 113
Grahm/Meus, Inc., 55
Greenboam & Casey, 76
Gresham, Smith & Partners, 158-159
Grid Formats, 20-21, 26-27
Griswold, Heckel & Kelly Assoc., Inc., 124

Hammel Green and Abrahamson, Inc., 58
Hansen Lind Meyer, 155
Harza Engineering Co., 182-183
Hatfield Halcomb Architects, 125
Havens & Emerson, Inc., 132
HDR Systems, Inc. 89
Helen M. Moran & Associates, 53
Hellmuth, Obata & Kassabaum, 164-165
Henderson Gantz Architects, 58
HLW 68-69
Holiday Cards, 78-81
Holt + Fatter, 72
Honeywell, 66
Howard Needles Tammen & Bergendoff, 186-187

Index, The Design Group of Laventhol & Horwath, 48, 92
Interspace Incorporated, 74
Intertech, 205
Invitations, 74-75
ISD Incorporated 73, 93

J.A. Jones Construction Co., 60
John Anderson, Inc. and Interior Environments, Inc., 95
John Wolcott Associates, Inc., 80

Kirkham Michael and Associates, 34-39
Kober Group, 110

Lea Group, 58
Leasing Brochures, 200-211
Lehrer/McGovern, Inc., 63
Logo, 6-7
Logotypes, 8-11
LZT Associates, Inc.,75,107

Magazines, 60-61
Master Plan, 34-35
Merger, 72
Michael Baker Corp., 176-177
Miles Treaster & Associates, 54
Minskoff Organization, 210
M.J. Brock, 138-139
Momograms, 4-5, 10-11
Morris/Aubry Architects, 68, 109
Myklebust Brockman Associates, Inc. 62

Nab Construction Corporation, 118
NBBJ Group, 80,83,87,88
Newsletters, 56-58
95 Wall Street, 208

O'Brien-Kreitzberg & Associates, 119
Office Design Associates/Shepard Martin, Inc., 24-25
Oliver Design Group, 26-27, 78
Opus Corporation, 190-191

Pan Am, 49
Parsons Brinkerhoff, 188-189
Partners Construction, 121
Peckham Guyton Albers & Viets, Inc.,105
Perez Limited, 85
Perkins & Will, 156-157
Pierce Goodwin Alexander, 48
Pop-Ups, 66-67
Post Buckley Schuh & Jernigan, Inc. 62
Post Cards, 51
Posters, 84-85
Preiss Breismeister, 81
Preface V
Professional Designs, Inc., 94
Project Pages, 24-25, 98-99, 102-103
Promotional Material Vocabulary, 45
Proposal Submittals, 86-87
Psomas & Associates, 77

Raymond Hansen Associates, 53
Rehler Vaughn Beaty & Koone, Inc., 17
Rockrose, 32,33
Rocky Mountain Geotechnical, Inc., 137
RTKL Associates, 79, 168-169
Rudolph/Libbe/Inc., 194-195

Schuchart & Associates, 117
Self Mailers, 54-55
780 Third Avenue, 209
Shope Reno Warton Associates, 98
Short and Ford Architects, 99
SHWC, Inc., 65, 77
Sikes Jennings Kelly, 84
Sketch Dummy, 22-23
Specialty Brochures 38-39
SMS Architects, 95
Specialty Brochures, 38-39
Spiral Bound, 95-97
Stephan Lepp Associates, 101
Stratton & Company, Inc., 142
Sugimura & Associates/Architects ,119
Sverdrup Corporation, 184-185
Swanke Hayden Connell Architects, 166-167
Symbols, 10-11
Syska & Hennessy, Inc. Engineers. 174-175

The Architects Collaborative, Inc., 102
The Aspen Club, 207
The Callison Partnership, 85, 154
The Earth Technology Corporation, 136
The Grad Partnership. 87
The Hillier Group, 130
The McElhanney Group, Ltd., 180-181
The Murphy Group, 84
The NBBJ Group, 80, 83, 87
The Ratcliff Architects, 57, 106
The Shops and Res. of Metro Dade Center, 205
The Stubbins Associates, Inc., 103
The Turner Corporation, 198-199
The Village Square at Stratton, 207
Thompson, Ventulett, Stainback & Assoc., Inc., 82, 162-163
Timber Court, 204
350 Hudson Street, 209
Tishman, 120

Trade Show, 88-89
Transamerica, 66
Trisha Wilson and Associates, 125
200 State Street, 204

Valley Ranch, 202
Van Dell and Associates, Inc., 135
Village Center, 203
Vitetta Group, 148-149

Walker Parking Consultants, 143
Washington Associates, Inc., 16
Williams & Burrows, Inc., 192-193
Williams Trebilcock Whitehead, 76, 80, 131
Wittenberg, Delony & Davidson, Inc., 126-127
Wood/Harbinger, Inc., 116
Woodward-Clyde Consultants, 115

Yearwood & Johnson, 56

Zimmer Gunsul Frasca Partnership, 152-153